**TECHNOLOGY, SUSTAINABILITY,
AND CULTURAL IDENTITY**

EDIZIONI PRESS

TECHNOLOGY, SUSTAINABILITY, AND CULTURAL IDENTITY

Foreword by Reed Kroloff
Architecture and Essays by Lawrence W. Speck
of PageSoutherlandPage

FOREWORD 6
BY REED KROLOFF

01/
ARCHITECTURE, GLOBALISM, AND LOCAL IDENTITY 9

ROUGH CREEK LODGE AND CONFERENCE CENTER 30

02/
A BROADER VIEW OF SUSTAINABILITY 39

ROBERT E. JOHNSON STATE OFFICE BUILDING 56
AUSTIN-BERGSTROM INTERNATIONAL AIRPORT 62

03/
THE PARADOX OF AMERICAN URBANISM 69

AUSTIN CONVENTION CENTER 88
COMPUTER SCIENCES CORPORATION 94
AUSTIN CITY LOFTS 100

04/
TECHNOLOGY AS A SOURCE OF BEAUTY 109

ADDITION TO SETON MEDICAL CENTER 124
AUSTIN CONVENTION CENTER EXPANSION 132

CREDITS 140

FOREWORD
BY REED KROLOFF

Twenty years ago, Lawrence Speck decided to launch a new collegiate architectural journal. At the time, American architecture enjoyed a robust academic publishing environment, with more than 15 magazines appearing regularly, including such venerable titles as *Perspecta*, *Oppositions*, and *Assemblage*. Speck, then as now, was teaching at the University of Texas at Austin, whose ambitious and increasingly wealthy architecture school was looking for ways to establish its scholarly credentials.

Larry Speck was a promising young academic with a small but active design practice, and was frustrated by what he viewed as the intellectual domination of the field by a tight group of elite east- and west-coast academic institutions and media outlets. Texas was booming, as were cities and states across the Sun Belt, and he felt the architecture and urbanism of those places was worthy of greater attention.

What to call this new publication? Speck's answer was telling not only for its clever play on geographic identity, but for the insight it offers into his architectural consciousness. *Center, a Journal for Architecture in America*, debuted with an issue entitled, "Architecture for the Emerging American City." It included articles by Denise Scott Brown and J.B. Jackson, icons of the architectural outsider school, and buildings by Ford Powell and Carson, Rob Quigley, Antoine Predock, and Speck himself, hardly household names in 1980's New York and Los Angeles. In rapid succession, Speck published two more issues, one on "20th Century Classicism," and another on "New Regionalism." *Center* would belong to no one camp, espouse no single ideology. Speck's instincts and timing were perfect. *Center* captured the pluralist ferment of contemporary architectural debate, won a wall-full of awards, and leveraged UT Austin's position considerably. Within five years, Speck was dean of the school.

Larry Speck's architecture is equally emphatic about not being emphatic. Like so many architects of his era, and like *Center*, he is not an ideologue: the Larry Speck who designed the historically inflected 1992 Austin Convention Center is the same Larry Speck who designed its modernist 2002 addition. Yet neither is he diffident: Speck's buildings may not look the same, but they share the same DNA.

Certainly he is a Regionalist. His buildings reflect the culture, ecology, and building traditions of central Texas (where most of his work is located). One can easily discern the area's German heritage in the House on Sunny Slope Drive and the Umlauf Sculpture Garden, with their understated shed forms and sturdy (and beautiful) limestone cladding. Similarly, Speck recalls local agricultural imagery at his Wimberly Ranch House and the Rough Creek Lodge and Conference Center. At the same time, his nuanced site strategies and sophisticated sections belie simple historicist readings. There's a modernist at work here. Aalto informs the sliding plan of Rough Creek, and the crisp detailing of Umlauf. Wimberly's vaulted roofs nod to Kahn's Kimbell Art Museum in Fort Worth, the precedent Speck finds most inspiring.

In short, Speck's work is difficult to pigeonhole, and, as his career has developed, so have his interests. His earlier projects, such as the *P/A* Award–winning Burnet Town Center, were more polemical, as befits a young academic practitioner. Now in mid-career, Speck has swapped leadership of a school of architecture for partnership in a large corporate practice. For the first time, he has the human and technical depth of field to extend his investigations into areas such as sustainability and high technology, and across programs of much greater size and complexity. Austin-Bergstrom International Airport is one of the larger recent projects, and Speck's hand is very much in evidence. The building is certainly sustainable in all of the more traditional senses of that word, with extensive daylighting, recycled materials, and local finishes throughout. But it is also sustainable culturally, with a soaring central court that transforms the facility from a transportation hub into a contemporary town square. Bands play there. Merchandise is sold. Local artists are on display. People actually like going to the airport. Imagine that.

At Austin-Bergstrom and other projects, Larry Speck's architecture offers a gentle manifesto on the authority of inclusiveness: architecture works best when people can find something of themselves in it. As with the center, Speck finds value in multiple viewpoints, and is unwilling to limit either his intellectual or his aesthetic horizons. Thus, while there is always rigor in his work, it never comes at the expense of reflection or revision. For the critic, that may make Lawrence Speck somewhat difficult to categorize. For the end user, it's easier: he's a good architect.

01 /

ARCHITECTURE, GLOBALISM, AND LOCAL CULTURAL IDENTITY

ARCHITECTURE, GLOBALIZATION, AND LOCAL CULTURAL IDENTITY

Architecture, at its best, embodies a society's consciousness about itself. More perhaps than the product of any other discipline—any of the arts, science, or technology—the artifacts which result from the act of building bespeak the character and aspirations of their makers. Architecture is our most timeless and quotidian means of expressing ourselves—of transferring ideas and values.

Buildings inherently and unavoidably document the everyday life of a culture. They are repositories of the patterns of activity, association, and movement of a society or people. Just as the Romans embedded in the city of Pompeii a treasure-trove of information that could be unearthed centuries later, so our current buildings and cities document contemporary life and attitudes. They do so in fundamental and everyday handling of land use, densities, building type development, plan configurations, and hierarchy of functions. But buildings also document contemporary life in more expressive ways that speak of the technological, social, and artistic values of a culture—its conservativism, adventuresomeness, intellectual capacity, emotional depth, farsightedness, ambition, inventiveness, etc.

For me, urban design and building design are pursuits which involve a mining of everyday life in a particular culture for perceptions and messages about its values, ambitions, and future. Architecture and city-building must address particulars of place and culture. They must be firmly rooted in the tangible realities of their situation—the history, climate, geography, economy, technology, human behavior, and cultural life of their place. When distracted by an inordinate preoccupation with artistry or style, architecture becomes flaccid, generic, and far less powerful as a cultural force.

It is easy to admire vernacular examples of environments deeply rooted in local culture. The poignant expressiveness of a New England village, a Rocky Mountain mining town, or a Southwest desert pueblo demonstrate real, tangible physical and societal diversity. Neighborhoods, towns, and regions that have held onto their authentic identities attract us by their toughness and individualism.

The little neighborhood of La Boca in Buenos Aires overtly proclaims its independent identity from the rest of its metropolitan region. Inextricably linked to the River Plate in a city that otherwise turns away from its river and port, "La Boca" ("the mouth of the river") grew up as a proud, working class Genovese neighborhood. Its ethnic roots (solidly Italian) as well as its industrial base (sailing, shipping, and boat-building and repair) created a strong identity early. Its development of a unique strain of tango and its breeding of one of the pre-eminent soccer teams in the world added layers of particularity. The physical environment of La Boca embodies and reinforces its cultural

01 Courtyard, "La Boca," Buenos Aires, Argentina
02 Streetscape, "La Boca"

identity. Inhabitants made their homes out of sheet metal just as if they were building ships out of the same materials. Impoverished locals begged excess paint from incoming ships and local repair shops, which they applied in a wild, unfettered way that expressed their raucous, exuberant way of living. The passion of their community, expressed in other ways as a romantic tango or a fanatical loyalty to local soccer deities, is palpable in its physical environment.

Modernity and Cultural Identity

One of the great contributions of modernism to the advancement of architecture came in its attentiveness to tangible realities and its rejection of *artistic style*. Louis Sullivan, the great American architectural pioneer, stated these early modern notions well at the beginning of the 20th century when he wrote, "Nothing more clearly reflects the status and the tendencies of a people than the character of its buildings. They are emanations of the people; they visualize for us the soul of our people."[1] Sullivan considered architecture to be a social and political medium capable of expressing ideals and values. He proclaimed, "To discuss architecture as the projected life of a people is a serious business. It removes architectural thought from a petty domain and places it where it belongs, an inseparable part of the history of civilization."[2]

The role Sullivan sought for architecture was not just a medium for reflecting longstanding culture, but an activist ingredient in seeking and provoking cultural progress. He was especially interested in architecture's role in helping to formulate American culture. He observed, "The spirit of democracy is a function seeking expression in organized social form.... Our self-imposed task is thus to seek out the simple: to find broad explanations, satisfying solutions, reliable answers to those questions which affect the health and growth of that democracy under whose banner we live and hope."[3] In particular, Sullivan sought out an architecture which expressed his own place—a "Chicagoey" architecture, one with "western frankness, directness—crudity if you will."[4] Modern architecture as Sullivan proclaimed it to his young colleague in *Kindergarten Chats* would be, "a beautiful, a sane, a logical, a human living art of your day; an art of and for democracy, an art of and for the American people of your time."[5]

These early modern notions found fruition in the work of many prominent 20th century architects but are perhaps most poignant in the work of designers who focused on eliciting a very strong sense of place in their work—architects like Frank Lloyd Wright, Alvar Aalto, Luis Barragan, and Louis Kahn. Informed by landscape, geography, climate, and local materials, and inspired by a careful observance of vernacular construction, these architects built, as Sullivan instructed, for the people and culture of their own time. Their work is fresh and inventive, but also is a very particular reflection of local culture.

03 Taliesin North, Spring Green, Wisconsin, Frank Lloyd Wright
04 Taliesin West, Scottsdale, Arizona, Frank Lloyd Wright

Frank Lloyd Wright and an Architecture of Locality

Wright learned Sullivan's lessons at the feet of the master himself. He, too, sought an American architecture and, particularly in his early career, an architecture for the emergent American Midwest. The Prairie Style house grew from its own native soil. Wright said, "I was born an American child of the ground and of space.... I loved the prairie by instinct."[6] His own home, Taliesin North, is all about embodying the rural life of Midwestern America. As Thomas Beeby has written in his meticulous analysis of the complex, Taliesin North is an extension of the geological and topographical forms of the region. Its dominant horizontality connects it to the broad, low sweep of the prairie. Its modest tower is a reflection of the unusual stone outcroppings which characterize this oddly unglaciated region.[7] The complex is made of the local ledge-stone in combination with the ubiquitous American wood frame. It learns from—but significantly advances—the rural farm vernacular of its environment.

This and other of Frank Lloyd Wright's houses of the first quarter of the 20th century were intended to portray the best of contemporary life and values of the American heartland. They were free-standing and individual. They exemplified self-sufficiency and democracy. They were unaffected, sincere, and honest. Wright's hero, Ralph Waldo Emerson, wrote essays on the values and ideals of his culture. Wright built embodiments of the values and ideals of his place and time.

The extent to which Wright relied on inspiration from the particulars of the place and culture where he was working to stimulate his inventiveness is evident in the extraordinary contrast between the home and studio he built for himself at Taliesin North in Wisconsin and the later compound he created for similar purposes at Taliesin West in Arizona. In the latter case, Wright is provoked by characteristics of the desert—its ruggedness, its flatness, its colors and textures. There is a radically different feeling than at Taliesin North. The Arizona colony even recalls literal vernacular building traditions of the region. The flat roofs, stepped terraces, and low, heavy walls made of stone gathered from the site all recall Pueblo construction. The buildings at Taliesin West are recessed into the ground at times like early settlers' dugouts in the region. Elsewhere, portions of the complex hover thin and light above the flatness of the desert like itinerate Native American tents. The ensemble is new, fresh, and vigorous—a real advance for Wright. But it is also bound timelessly to its place.

At almost every level, Wright's projects in Wisconsin and Arizona reflect a poignant sensitivity to local building and culture. Taliesin West is all about shade from the sun, catching breezes, being outdoors much of the time, and reveling in the starkness and ruggedness inherent in the landscape. There is a lifestyle and body of cultural values embedded in the buildings that Wright adopted from the desert and its inhabitants. Taliesin North, by contrast, is all about drawing the precious winter

05 Villa Mairea, Noormarku, Finland, Alvar Aalto
06 Villa Mairea—living room

light indoors, creating sunny, outdoor spaces protected against the north winds, and enjoying the soft green character of the landscape during warmer seasons of the year. There is a gentleness and civility that is very rural and Midwestern. The complex is clearly rooted in its agricultural culture and landscape with functional outbuildings oriented to crop cultivation.

Wright's modernism was not about a universal set of forms or techniques. It was not just a new style. It was about responding to the tangible realities of a situation. It was, in Sullivan's words, "serious business" which required being "sane" and "logical" in order to create "satisfying solutions" and "reliable answers." Wright, at his best, was capable of addressing a place like Wisconsin or Arizona and making magic of the elements he found there—elements that might have been overlooked or undervalued by other designers for generations. Whereas many architects would have written off both regions as having little or no substantial form-giving impetus to relate to, Wright discovered the latent potential even in contexts that seemed to be unpromising.

Alvar Aalto—Capturing a Sense of Finland

Wright's talent is akin to something Alvar Aalto noted as "the gift of seeing the beautiful in everything."[8] Like Wright, Aalto often built in places that did not possess a strong, cohesive architectural heritage. Yet he felt a genuine desire to draw on the very best qualities he could discover in whatever place he was working. In his earliest projects, this inclination took the form of capable reliance on both the Finnish rural vernacular of simple board and batten volumes and the Leningrad-inspired Neo-Classical vocabulary established by C. L. Engel in the early 19th century in Finland.

By mid-career, however, Aalto was capable of a much richer and more appropriate regional expression that brought together more synoptically "the beautiful in everything" in Finland. The Villa Mairea of 1939 is, like Wright's work at Taliesin North and Taliesin West, an extraordinary response to the tangible realities of its place and a landmark work of modern architecture. The spacious rural home built for a prominent Finnish industrialist family is part Nordic sod-roofed hut, part vernacular log cabin with gutters hewn from tree trunks, part reinterpreted board-and-batten-clad farmhouse. But it is also part Scandinavian Functionalism that had developed quickly in the decade of the 1930s and part new industrial Finland, with its rapidly developing wood-products and ceramics manufacturing.

The Villa Mairea is crafty, rugged, and relaxed like the landscape and building traditions of Finland, but it is also orderly, clean, and precise like the mindset and culture of Finland. The regionalism in this instance is not a one-liner. It draws on the climate, the shapes in local topography, and the colors and textures of the landscape, as well as on lifestyles and social customs. The integration of sauna and dipping

07 Egerström House, Mexico City, Mexico, Luis Barragan

pool in the heart of the building ensemble, as well as the creation of an almost-enclosed court around which various functions gather, ground the daily life of the home firmly in Finland. Aalto invented a fresh new regionalism—not a style, but a carefully crafted sensibility for building in his place.

Luis Barragan—A Modern Mexican Expression

Concurrent with Alvar Aalto's development of his regional sensibility in Finland, Luis Barragan was pursuing a similar search for an appropriate regional expression for a very different landscape, climate, and culture in Mexico. His accomplishments are comparable to Aalto's. Barragan loved Mexico in much the same way Aalto loved Finland. He was devotedly Mexican, refusing to work elsewhere even when tantalizing commissions were offered. He knew Mexican vernacular building instinctively from his childhood and deeply felt its strength and potency. He had an educated appreciation for Mexican religion, painting, and literature. He tapped these deep-seated resources in creating his extraordinary architecture.

Barragan's memorable fountains evoke images of the wood aqueducts that spanned the courtyards and passageways of the village of Mazamitla that he knew from his childhood. His walls draw on the ubiquitous Mexican traditions of masonry building in addition to the common Mexican convention of creating rigorous spatial distinctions between public and private realms. His signature vibrant colors are not only rooted in the brilliant hues of the everyday urban scene in so much of Mexico, but are also refined by the studious experimentation with color of Mexican artists like Rufino Tamayo and Barragan's close friend and collaborator, Chucho Reyes. Even the construction technology of Barragan's buildings frequently relies on methods of traditional construction common in vernacular pueblos and haciendas.

But Barragan's work is also tied to the 20th-century fervor in Mexico for modernity. Barragan, like most of his contemporaries in the 1930s, had a subsuming love affair with Le Corbusier's emerging vocabulary of the period. His later work validates the legitimacy of these early influences, but puts them in proper perspective. Barragan's modernism welcomes the evolution that comes of incorporating the interests of each new generation into longstanding cultural traditions.

Louis I. Kahn—Building with a Strong Sense of Place

The closest parallel to this fresh new regionalism in the modern era in the United States is the late work of Louis I. Kahn. Two of the last buildings he finished before his death make the point very strongly. His Exeter Library in New Hampshire is chaste, subtle, and reserved on the outside—a quintessential New England building. It is deferential, even acquiescent, in its campus context. Its simple stereometric volume pierced by a regular pattern of repetitive wood windows, its taut red

08 Exeter Library, Exeter, New Hampshire, Louis I. Kahn
09 Kimbell Museum, Fort Worth, Texas, Louis I. Kahn

brick envelope, its vertical hierarchy from massive and robust at the base to delicate and diminished at the top all position it easily within the longstanding building traditions of New England. The Kimbell Art Museum in Fort Worth, Texas, of exactly the same period, is as strikingly different from Exeter Library as, in fact, north central Texas is different from southeastern New Hampshire. Again, Kahn relies heavily in the Kimbell Museum on regional inspiration. The broad horizontality of the north Texas prairie is reflected in the low, flat character of the building. At the front is the familiar Texas porch that also serves to introduce and explain the building's carefully articulated spatial and structural system.

To anyone familiar with the area, the long concrete vaults that are the units from which the building is composed draw an easy parallel to the cylindrical concrete grain elevators that are landmarks throughout the city and the surrounding region. Kay Kimbell, who was the benefactor of the museum, made his early fortune, in fact, in grain. But, horizontal as these features are, they are a closer parallel still to the bow-topped stock barns that stand in regular ranks just a few blocks from the museum and are typical of the galvanized metal industrial vernacular of the region.

Kahn is not just sampling eclectically from forms he observed along the highway driving from the airport. He demonstrates a real and profound feeling for Texas—particularly in his handling of light, color, and texture. The tawny tans and grays of the building are reminiscent of the colors of the native landscape, with its limestone substrata and parched grasses. The tactile quality of the building's materials gain even greater character when placed in deep relief by the harsh Texas sun. Everywhere, light and shadow are modulated with great understanding and finesse. The building is truly resonant in its place.

All four of these architects—Wright, Aalto, Barragan, and Kahn—stand among the most significant innovators and form-givers of the modern era. The inspiration for that innovation and form-making is significantly from local conditions and culture. Our understanding of that origin is frequently clouded by the fact that architecture is so often delivered to us through media severed from its cultural ties. It is hard to truly understand the rootedness of Wright, Aalto, Barragan, and Kahn by a handful of closely cropped images with little real knowledge of Wisconsin, Arizona, Finland, Mexico, New England, or Texas. The reality is most clear when you are there, experiencing a building as a tangible artifact in and of its culture.

A Contemporary Practice Emphasizing Locality

In our own work over the last 25 years, we have made a concerted effort to research and understand the culture and character of the places where we work and to use that knowledge to inspire design directions. We believe that varied local cultures are indeed alive and thriving across

10 Freeway Park, Seattle, Washington
11 Paseo del Rio, San Antonio, Texas

the United States and all over the world. Though conventional wisdom seems to say that advances in transportation, communication, and media in the 20th century have tended to homogenize culture, our observations tend to indicate the opposite in some ways.

As people become more mobile they are able to select what city or region they want to inhabit. People who love the outdoors and active recreation choose places like Colorado or Wyoming. People who like theater and opera flock to New York or Los Angeles. Individuals who thrive on following big-league sports teams choose cities where that is a focus. Those with an intellectual bent may prefer a college town. The point is that, more and more, people are not stuck living where they were born but have options to congregate with like-minded people in communities that really have something in common—that can create a meaningful culture out of shared identity.

Indianapolis, which was once dubbed "India-no-place," has, in the last decades, developed a strong cultural identity around sports and athletics. Nashville, once a very sleepy southern town, has recently emerged as a powerful center for the entertainment and recording industry grown from roots in country and western music. Portland, Miami, Tucson, Raleigh-Durham, Salt Lake City, Boulder, Austin, and Seattle have all created strong new cultural identities for themselves in recent years and have joined San Francisco, New Orleans, San Antonio, Boston, and New York as American cities with a real uniqueness and local character. Young people flock to these cities with a very strong sense of who they are and what they want in a community to sustain them. They are rewarded by finding economic, social, political, lifestyle, and entertainment options that suit them.

Unfortunately, to date, the physical environments of American cities have failed to reflect this genuine diversity of culture and context as much as they might have. Generic stylistic trends seem to formulate architectural character more than tangible realities such as local climate, geography, technology, values, and lifestyles. In our practice we have tried to adopt modern principles of responsive, analytical, logical, and particular design to create architecture firmly rooted in its place and culture.

The physical environment can be a potent mechanism for investing identity and a sense of rootedness. The German term *heimat*, sometimes inadequately translated into English with words like "homeplace" or "homeland," describes a notion of attachment to a place—a psychological bond which can be both nurturing and stimulating. If local cultural identities built around shared values and common predilections can be reinforced by this kind of psychological bond to a meaningful physical environment, a true sense of community can grow.

Much has been written about the negative effects of a kind of neutralizing globalization that has occurred in both political and economic realms in recent decades. Rebellion against loss of local ethnicity, religion, cultural practices, and beliefs has been swift and extreme in

12 Live oak, Central Texas
13 Landscape, Central Texas

many parts of the world. But even outside the most dramatic of these rebellions, there has been a widespread and poignant sense of loss in instances where local food, local music, and local commerce, for instance, get obliterated by the massive corporate machines of globalization. Eric Schlosser, in his book *Fast Food Nation*, makes an excellent case for resistance to these forces. He makes it clear that it is not just the bland, superficially-appealing-but-unhealthy nature of the food that is destructive in the industries he investigates, but also the neutralizing impact on physical environments. Resistance can seem a daunting challenge, but Schlosser is encouraging in his most recent edition, observing some signs of very recent progress.

Architects have a responsibility to advocate for the creation of authentic and meaningful environments which support the cultures they serve. Doing so requires developing sensitivities and skills in reading and understanding particular cultures and places in much more than just a superficial manner. This may take the form of a long-term commitment to a place and culture as in the case of Luis Barragan's life-long love affair with some particular aspects of Mexican life. But it can also take the form of the gimlet-eyed outsider's view as in the case of Louis Kahn's extraordinary perceptiveness about North Texas in the case of the Kimbell Museum.

We have been fortunate to be participants in the growing of an exciting local culture in Austin and Central Texas over the last several decades. It has been an exciting place to be and to practice architecture as a medium of cultural expression. We have fed on the insights of local writers, artists, and musicians. We have been activists in common causes with local politicians, arts advocates, and environmentalists. Our architecture is a subset of a larger commitment to helping our local culture become strong, independent, and assertive, propelling it to influence in the larger world.

In recent projects outside our region and our state we have sought to apply the sensibilities we learned at home to gain trenchant insight into the cultures of other places. This is a natural evolutionary step. Appreciating and valuing one's own local culture does not mean being stymied or restricted by it. On the contrary, it opens one's eyes to the nuances and particularities of any other place when traveling or working in a new environment.

Four Houses in Central Texas
In America, historically, there is perhaps no building type so emblematic of our local cultures as the single-family house. Though it has generated some significant urban challenges, it has also contributed a way of life that can be warm, rich, individual, and closely connected to nature. In Central Texas, where the natural environment has great strength and beauty, the possibilities for a single-family house inextricably interwoven with the landscape are particularly powerful. There are, among the

14 Limestone wall, Central Texas

current generation inhabiting this region, a significant subset who appreciate deeply the land and its beauty—who understand how a whole lifestyle can be built around an integration of everyday activities inside and outside in this bold, varied terrain and climate.

As disparate sources as Lady Bird Johnson and the "hippie" counterculture movement of the 1960s began focusing attention on the value of the Central Texas landscape in the 1970s. Seen as an antidote to the rampant generic urbanization of Houston or Dallas, Austin and the region around it seemed to offer an alternative based in a genuinely divergent lifestyle as well as in the hills, creeks, bluffs, draws, and vegetation of its unusual geological circumstance.

These four houses were all done for clients who loved the Central Texas landscape and who drew from it and from other cultural sources a way of living that was particular to its place. In each instance we worked together to create a house that would not be generic or replicable elsewhere. We tried to make real Central Texas houses that would seem odd or peculiar out of their context but would fit comfortably and amiably in their place. In their own modest way each of these houses protests the globalization of lifestyle, art, and economy. They reject mass marketing and the primacy of "resale value." (Three of the four original clients still live in the houses 20 years later.) Outside their role in accommodating the lives of their inhabitants, these houses seek to explore what is interesting and distinguishing about this place and this culture. They revel in the crisp clear light of Central Texas with rich texture, color, and articulation. They tame the hot Texas sun with deep shady recesses varied logically and analytically according to orientation. They are generally strung linearly one-room-deep, perpendicular to prevailing breezes to optimize natural ventilation.

They are constructed of a mixture of old and new materials—mostly locally produced. The ubiquitous Central Texas limestone is used generously, but not in a thin veneer as is predominantly the case in current practice. Stone walls here are built of two wyeths of masonry with an airspace filled with insulation between to create high thermal mass which contributes to keeping interiors cooler during the long Texas summers. Reinforced concrete is often integrated with the stone when a more demanding structural situation requires. Louis Kahn called concrete "liquid stone," and it is used here with the limestone as a sister masonry material. Cement plaster or cement panels extend this masonry vocabulary to lighter, thinner walls.

Galvanized sheet metal and, occasionally, wood siding round out the palette of exterior materials. The thinness of the metal allows it to reflect heat without transferring it into the building. Its ductile nature enables it to take looser, freer forms than other, more rigid and brittle materials. Wood siding is used in a straightforward manner as a sensible sheathing for balloon frame construction. It is treated to weather naturally to a dark gray-brown.

15 Burnet Ranch House, Burnet, Texas
16 Burnet Ranch House—plan

Because stone, concrete, cement plaster, sheet metal, and wood have a history of sophisticated use in the region, there are craftspersons who can handle them skillfully. None of these four houses had large budgets. They were generally not built by the "premier contractors" who specialize in high-end custom homes. They were built simply and economically utilizing skills available in the local workforce. These houses were intended to be a kind of modern vernacular—rooted, particular, responsive, and accessible.

The Burnet Ranch House

The Burnet Ranch House was designed for a young family of four on a working sheep and goat ranch outside Burnet, Texas. The parents were well-educated individuals who had traveled widely and lived in many different parts of the world. Their decision to live in this rural setting bucked conventional wisdom for sophisticated people of their generation. Their love of the land and the nurture they drew from it, however, compensated for the loss of urban culture. They had owned the ranch for several years before building a house there, placing the establishment of barns and service infrastructure as well as the planting of a pecan grove as a higher priority. The site reserved for the house had commanding views of the ranch in every direction.

The house is very much about being on the ranch—enveloped by its peacefulness and reliable rhythms of daily life. More a compound than a building, the house is an agglomeration of five small interconnected pieces—each taking its character from the particular functions within, and the orientation to external views, sun, and breezes. A central wood-clad, two-story element marks the terminus of the winding ranch road approach and enfronts a pond below. It houses the entry and a very large kitchen designed as much for conversation as for cooking below and the two children's bedrooms above. Its deep double-decker porch catches southeast breezes off the pond and extends all of its functions into the outdoors.

To the south of the central element are a stone volume (housing a washroom for cleanup coming back from the barn on the lower floor and playroom off the children's porch above) as well as a carport which can double as a place for outdoor entertaining. To the west of the central pavilion is a tall single-story volume which houses a single grand room designed to be the internal focus of the house. Views from this wing are oriented north through a deep summer porch, toward the rugged hills of the ranch's grazing lands, and south through a gallery and winter porch, which opens onto an outdoor court. (The house has almost the same number of square feet devoted to porches as to indoor spaces.) The fifth and westernmost building element is a wood-clad volume housing the master bedroom and bath. Its angle closes the outdoor court spatially and shields it from the western sun.

The Burnet Ranch House grows out of its local culture and place but it's also very cognizant of the culture in its time. An article published

17 Ranch House, Wimberley, Texas
18 Ranch House—section
19 Ranch House

in *Progressive Architecture* shortly after its completion observed: "[the Burnet Ranch House's] deliberate dissonance, a helter-skelter jamming together of things... bespeaks the late 20th century while acknowledging the particularities of function and site." The building weaves together issues of land conservation, sustenance of rural lifestyle, and sustainability, which have not only strong local resonance but broader implications as well.

The Ranch House in Wimberley, Texas

The Ranch House in Wimberley, Texas, is situated, like the Burnet Ranch House, on a plat of several hundred acres. The house site is on a high bluff shaped like a bowl above the Blanco River—one of the most scenic waterways in the region. It creates an edge between the high flatter part of the ranch to the north and the soft green gorge of the river, lined with huge cypress and live oak trees, to the south. The rooms of the house are strung in a bow to embrace the bowl-shape of the land and to orient views directly toward the river.

Two barrel-vaulted bedroom suites cant slightly southeast to catch prevailing breezes more directly. They each combine a bathroom, closet, and sitting room, dominated by a stone fireplace below and with a sleeping loft above. The sitting rooms spill directly out to a large shared terrace shaded by existing trees. From the sleeping area there is a powerful connection both horizontally and vertically from river, to bluff, to terrace, to sitting room, to loft, to ranchland. The vertical volume of the suites draws cool air up from the shady riverbed and ventilates it out the top of the vault so that the rooms are generally pleasant without air conditioning even on a very hot day. Metal eyebrows cover all operable windows so they can be left open even when it rains.

At the opposite end of the house from the bedroom suites is a very large kitchen designed for several people cooking together and for occasionally feeding large groups. (For a while the kitchen was home to a notable local cooking school.) It is a dramatically tall space with a ceiling that slopes up in three directions to the exterior wall. The ceiling acts both as a light deflector, bringing soft generous light to work counters from high clerestory windows, and as a means of directing warm air from cooking up and out the operable openings. Just above counter height are generous windows that give views down to the river and direct cool breezes up from the river to counteract the heat generated in the kitchen.

Nestled between the taller bedroom suites at one end and the kitchen at the other is a long lower room housing living and dining functions. A continuous terrace links all three volumes on the river side, shaded by the canopy of live oaks as well as deep roof overhangs.

The local culture here is somewhat different than that of the Burnet Ranch House. Wimberley has become a kind of resort community with weekend homes for many Houstonians. Though this particular

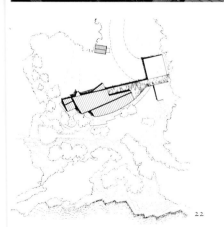

20 Lake Travis House, Lake Travis, Texas
21 Lake Travis House—living room
22 Lake Travis House—site plan

house has been the primary residence for its owners, it participates in a different kind of interaction with functions of the landscape. The ranch owners here had already begun to clear the land of invasive non-native vegetation and to restore indigenous grasses before the house was built. The scenic and ecological nature of the land is more important than its productivity. The Wimberley Ranch House revels in that scenic and experiential confluence with the land. It is about enjoying a well-prepared meal under the dappled light of a spring afternoon and feeling the soft breezes off the river caress your skin. It is about sophisticated sociability—having guests and entertaining. It is about good food, good wine, good conversation, and a setting that amplifies these everyday pleasures. The looseness, responsiveness, and authenticity of the architecture is an embodiment of a lifestyle and set of values generated by a memorable landscape and place.

Lake Travis House

The Lake Travis House is a step more urban than the two previous ranch houses though it still sits on several acres of land oddly left out of the resort subdivisions that surround most of the man-made lake it enfronts. Perched on a bluff 100 feet above the water at Arkansas Bend in Lake Travis, the site commands extraordinary views to the southwest, west, and north. Acknowledging both the variety and panorama of the view potential, the west side of the house takes the form of a sweeping arc while the north views are captured in vertical, stepping rectilinear forms. Orientations of wall shapes and fenestration are matched to specific characteristics of various vistas. With so much orientation to the west, sun control is a dominant design factor. Detached concrete piers, horizontal wood trellises, deep awnings, and concealed automatic rolling shutters on the exterior provide very thorough control through the full range of seasons.

Like the two ranch houses, the Lake Travis House derives much of its identity from the landscape, climate, and other physical characteristics of its locale. Like the others as well, there is a social dimension that reflects the nature and values of the client as a subset of local culture. In this case, the couple who were clients for the house were early predictors for a kind of Central Texas habitué that would become prominent over the following two decades. The husband is a software consultant and the wife an environmental activist particularly interested in native plants. The couple and their two young sons could have lived pretty much anywhere since neither of them had jobs that required being in a particular location. They chose to live in Central Texas because they felt an empathy with the place, and, like thousands of others, their move to Austin reinforced the extant local culture.

The spatial organization of the Lake Travis House reinforces a lifestyle this family chose explicitly. They had purchased the land several years before they built the house, living in a very inadequate existing

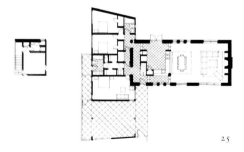

23 Lake Travis House—west face
24 Lake Travis House—detail
25 House on Sunny Slope Drive, Austin, Texas—plan

structure until they could afford to build. During that period they had studied every nuance of the land—where the sunset was most glorious at what time of year, which plants flowered and when, migration patterns of wildlife, etc. The house and its inhabitation of the land needed to reinforce the pleasure they gained from being in this place. Like the ranch houses, views and outdoor spaces became as important in terms of everyday function as interior spaces.

But there is also a pattern created in the interior rooms that reflected a particular emerging local culture. For this family, and many others like them moving to Central Texas, living, working, and raising children did not have discrete boundaries. Work hours and leisure time were not defined as separate. Much professional activity occurred at home—often in conjunction with child care. Domestic chores like laundry and cooking needed to be made easy and integral to other activities. The lower floor of the house responds to these characteristics by becoming an open series of spaces oriented to terraces and lake view, which have, at the center, a focus on cooking and eating. Child-centered spaces anchor the south end of the lower floor, while work-centered spaces anchor the north end. Upstairs, bathing (including an outdoor shower) and laundry become the hub in the center, with children's bedrooms off a huge outdoor play area on the south end and master bedroom oriented to additional views to the north.

The Lake Travis House embodies, in a direct, unaffected way, the longstanding physical and emerging social culture of its place. It nestles gently into its site, shrouded by existing vegetation even though it sits in a prominent location high above the lake. The life lived within and around the house becomes a natural extension of life that has occurred there before. An old log hut remains comfortably near the house. A concrete bomb shelter from the 1950s is integrated into the carport and terraces. The house is a contemporary extension of the land and its occupation expressed logically, sensibly, and modestly.

House on Sunny Slope Drive

The fourth of these Central Texas houses completes a continuum from the fully agrarian Burnet Ranch House to a location on a small lot in a dense in-town community in Austin. The House on Sunny Slope Drive is located on a 75- by 90-foot lot in Tarrytown, a well-established mixed-use neighborhood just west of downtown and the University of Texas. The 2,300-square-foot house was placed deep on the back of the lot in order to combine the front yard (mandated by a required 25-foot setback) and what would have been a narrow backyard, into a single, more usable outdoor space. This larger front garden is shrouded from the street by a six-foot fence made of the same materials as the house.

Because the neighborhood has many different kinds and sizes of houses from many different eras, there is less consistency in terms of how houses address the street than in "built-at-once" suburbs. Even on

26 House on Sunny Slope Drive—view from the street
27 House on Sunny Slope Drive—south-facing wall
28 House on Sunny Slope Drive—living room

little one-block-long Sunny Slope Drive there were wooden bungalows with big front porches, 1950s split-level ranchburgers, a two-story near-colonial, and even a very sophisticated 1960s modern brick and glass house surrounded by walled gardens. The wonderful, eclectic permissiveness of Tarrytown genuinely reflects the rich diversity of its inhabitants. There are cultural messages here that are powerful indicators of Austin as a community. The mixture of large houses and small houses, the intermingling of fairly well-off families with households of modest means, the integration of young families with older people who have spent most of their lives in the neighborhood, all contribute to making Tarrytown, along with six to eight other in-town neighborhoods, quintessentially expressive of Austin.

The decision to create a private garden in the front of the house fit naturally and comfortably into the character of the neighborhood providing, now that landscaping has grown fully, a kind of green interlude on the street. The house itself pretty much disappears. The interlude reinforces the freedom and variety of the street and the sense of diversity of the neighborhood. Its independence reflects one of the very best aspects of Austin culture—an appreciation for individuality and loose rules. (This is this same set of values that has fueled Austin's eclectic live-music scene.)

The primary wing of the house, tucked to the rear of the lot, is a simple rectangular volume made of 18-inch-thick stone walls housing living, dining, and kitchen functions as well as a small loft study. It is one big open room inside. The stone walls give a strong sense of privacy and protection, but large, strategically located openings give a sense of being intimately connected to the leafy understory of the trees outdoors as well. Even here in town, there is a sense of the toughness and ruggedness of Central Texas contrasted with the softness of shade and green. The openness of the room and its generous light as well as its textures, colors, and materials all tie it clearly to its place.

The south-facing front wall has a deep overhang created by an extension of the galvanized metal roof framed on exposed steel struts. This heroically scaled shading device works with the high thermal mass of the room inside its insulation to keep the space cool in the summer. Operable windows oriented for access to prevailing breezes and cross-ventilation further contribute to providing a temperate thermal environment with a minimum of mechanical support.

A secondary wing hugs the west side of the lot protecting the garden from late afternoon summer sun. It houses three bedrooms, two baths (one with an outdoor shower), and a "carporch." The latter space is both a generous covered patio extending the outdoor room of the garden for entertaining and recreation, as well as a place to store the car under shelter when desired. Bringing the car into the house in a frank, unabashed way is natural and sensible in the context of the car culture inherent in any new American city. The car here is a comfortable, compatible element as integral as any other domestic machine.

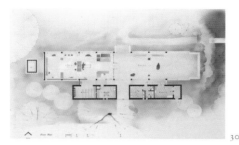

29 Umlauf Sculpture Garden, Austin, Texas—south face
30 Umlauf Sculpture Garden—plan
31 Umlauf Sculpture Garden—north face

The secondary wing and garden wall are made of horizontal strips of cement board—an industrial product more commonly used as form-liner in construction than as a finish material. Durable and easy to cut and work, the material has weathered beautifully through the years, creating a compatible, but differentiated character for the more modest parts of the house in comparison to the great stone room. The composite is, as *Architectural Record* noted at the time of the house's completion, a building with "respect for Texas tradition" but also "a low-key modern house." It is comfortably both of its place and time.

Umlauf Sculpture Garden

Located at the edge of Zilker Park, a sprawling urban oasis near downtown Austin, the Umlauf Sculpture Garden is a large, naturally landscaped precinct housing the work of sculptor Charles Umlauf. Probably the best-known Texas artist of his generation, Umlauf did large-scale works, primarily in bronze, which are well-suited to outdoor display. But he also worked in stone and wood and at a smaller scale which required an indoor gallery. The site sits at the base of a 50-foot-tall limestone bluff atop which is the home and studio where the sculptor lived and worked for more than 40 years.

A building to accommodate the gallery and other support functions was located on a degraded portion of the site, which had been used for a landfill. This created an opportunity for site-repair as well as a chance to preserve the more scenic, wooded areas for gardens. The building was strung long and thin across the short dimension of the plat in an effort to define a small parking area at the south end of the site, completely shrouded from the gardens by the building. The gardens at the north end of the site would then be contained by the building on one end and the bluff with the artist's compound on top at the other end.

The gallery is composed of two distinct volumes. On the south side, a long, thin stone element with a wide sallyport puncturing its center creates a dramatic gateway leading from the parking area to the gardens. This volume contains a small media room, library, rest rooms, and kitchen which require minimal exposure and act as "servant" spaces for the rest of the building. On the north side, an even longer open-frame volume houses both indoor and outdoor gallery spaces. A grand, generous porch comprises half this volume, receiving the sallyport on the south and opening out to the broad expanse of the gardens to the north.

As with the four houses, the Umlauf Sculpture Garden is fundamentally shaped, both in site-planning and building character, by qualities of the natural landscape. The solid limestone bar of the building parallels the dramatic limestone bluff behind it. The open frame of the gallery pavilion has similar qualities of space and light and vertical articulation to the densely canopied woods where the gardens are located. The north-facing glass wall of the interior gallery not only provides abundant, soft light for the sculpture within, but also connects

32 Umlauf Sculpture Garden—detail
33 Sulzer Orthopedics Facility, Austin, Texas

the interior visual experience strongly to the landscape. The entire experience of the place is about being in the garden.

Consistent with this approach, the expression of the building itself is very quiet and understated. The volumes are generous and the north light is rich and warm. Its materials—limestone, weathered wood, clear anodized aluminum window frames, and a galvanized metal roof—are sympathetic with the colors, textures, and scale of the trees, the bluff, the tiny streambed, and the sky. The transition from road, to parking, to sallyport, to porch, to terrace, to gardens, to bluff is a slow bleed from man-made to natural and from everyday urban life to a natural oasis in the heart of the city.

For many people in Austin, the Umlauf Sculpture Garden is in an iconic Austin place. Just a stone's throw from Barton Springs Pool—the ultimate iconic Austin place—it seeks to share the same appreciation for the crisp air, big sky, hot sun, deep shade, lush green vegetation, and cool breezes that are so palpable in the pool precinct. The garden, and, in particular, the great porch in the gallery building, have been used for a wide variety of events that are important to remember. Political occasions, social events, weddings, and even a number of memorial services have been held on the porch because of its strong identity and sense of place.

There is a set of values, an attitude toward life, and a way of living in this instance that is particular to the local culture. The role of architecture here is not to create an art object or a representational image. It is rather to define a place where ritual and occupation become the focus—where life and culture are embodied in the character of the environment. It is the feeling and spirit of the place that are its most important qualities.

Sulzer Orthopedics Research, Operations, and Production Facility
Essential to the "new" culture of Central Texas which has been developing over the last several decades has been the growth of high technology employers who have been drawn to the region by a well-educated workforce and a desirable "quality of life." One such company is Sulzer Orthopedics, a biotechnology firm with Swiss ownership. Though this kind of semi-industrial facility is seldom thought to have significant "architectural potential," this was a very exciting project for us. It offered an opportunity to take a generic, everyday building type which occurs on the outskirts of most American cities and see if we could "regionalize" it to Central Texas.

Sulzer, an international company relatively new to Austin, was anxious to make a building that "fit" there. They decried the image of the faceless multi-national corporation and relished the notion of a workplace that reflected the values of their local employees and the culture they were a part of. This was a chance to demonstrate that a corporation could participate in a global economy without contributing to the degradation of local culture and identity, which has been the

34 Sulzer Orthopedics Facility—company street
35 Sulzer Orthopedics Facility—dining and meeting pavilion
36 Sulzer Orthopedics Facility—site plan

byproduct of so much globalization. The Swiss owners, perhaps because of their own tiny country's proud sense of identity, were very keen to build this large new facility with Central Texas clearly in mind.

The site chosen for the new building was in a new industrial park northeast of the city. Thankfully, though one road had been installed by a developer with great destruction to the landscape, Sulzer was able to acquire a large portion of the park for both immediate and expansion needs and to take it out of the developer's control. The western half of the tract they acquired had some substantial trees—especially at the southern end. The eastern half was relatively flat and bald. A commitment was made to leave most of the western half as a preserve, with only a bit of executive and visitor parking woven among the trees, and to concentrate both immediate and future development where the landscape had already been cleared and grazed as pasture land.

The building was placed at a diagonal to the cardinal points allowing both northwest and southwest faces to look into the preserve. An L-shaped, two-story research and operations compound was placed on this side of the building. A one-story square-shaped production facility was nestled in the elbow of the L-shaped wings. A tall light-filled "company street" created a bridge between the production facility on one side and research and operations spaces on the other. The goal was to create a common space where blue-collar and white-collar worlds mixed and where sales and management people could interact informally and naturally with scientists and production workers.

At the intersection of the two legs of the "company street," a large cafeteria and meeting space provided a landmark and anchor. Its views out into the preserve in two directions and its double-height space linking the various parts of the company vertically as well as horizontally give it a distinctiveness that is almost civic.

The organizational patterns inside the Sulzer Orthopedics building are a blend of the way the company operated in general and a particular sense of how their company might best operate in the specific work context of Central Texas. An elaborate polling and interview process revealed a more integrated, less hierarchical attitude among the Austin group which found physical expression in the tight clustering of research, operations, and production together, and in the idea of the company street. A great desire for connections to outdoors for breaks and lunch resulted in a much more permeable building than usual with strong links especially to the most scenic parts of the site off the southwest face.

The architectural expression of the Sulzer Orthopedics building reflects both its internal organization and a strong responsiveness to the surrounding landscape. The two-story research and operations wings are simple, flexible loft spaces built of deep stone piers and concrete spandrels. They are repetitive and straightforward—tectonically simple and logical. The voids created by the piers and spandrels become very large 12-foot by 12-foot apertures giving dramatic vistas out to the

37 Sulzer Orthopedics Facility—research and operations wing
38 Rough Creek Lodge and Conference Center, Glen Rose, Texas

preserve. There is nothing cloying or sentimental about these volumes. They are clearly industrial and technological. And yet there is a clear empathy with the land. The thicket of live oaks, native grasses, and prickly pears feel comfortable and compatible at their edges.

The double-height dining and meeting pavilion at the crossroads of the company streets contrasts with the research and operations wings by being much lighter and more delicate. A glittering jewel-box at night, it depicts the company as a whole rather than an agglomeration of parts. The single-story square of the production facility is sheathed in a carefully detailed corrugated metal skin befitting its role as industrial workhorse. The two legs of the company street, each with a slightly different section to avoid monotony, are more material-neutral so they can bridge together the other more material-specific building parts. They revel in the rhythm of movement and in the generous top-lighting from clerestory windows.

Sulzer Orthopedics demonstrates an attitude by which the wastelands of generic office and industrial parks which ring our cities could become rich, particular workplaces that contribute to local identity and sense of place. The composite effects of dozens of such facilities could be the retention of general native landscape character, a compatibility of buildings and landscape and far greater visual and environmental amenity in work environments. The budget for Sulzer Orthopedics was industry standard. Additional design character and environmental assets had to come from very simple, economical means. It is precisely in everyday economy-driven situations like this that design has great power to make a significant difference in our cities and in our lives.

Rough Creek Lodge and Conference Center

Our project for Rough Creek Lodge in Glen Rose, Texas, is a good example of a building substantially formulated as a response to natural and geographical conditions such as geology, landscape character, climate, materials characteristics, and environmental compatibility. The specific site for the complex was chosen as the most advantageous location for the required function on an 11,000-acre ranch in north-central Texas. The function to be accommodated was an executive retreat center—a place for respite and contemplation, but only an hour and a half drive from a major intercontinental airport.

The lodge occupies a high ridge with commanding views to a Y-shaped lake in the foreground and very long distant views enhanced by taller, shapely peaks beyond. There is a grove of now-mature live oak trees on the site, planted several generations ago. The buildings are strung long and low along the ridge, hugging the contours of the site. They are nestled in and around the trees so as to seem like a natural outgrowth of the ridge.

The view side of the site faces mostly north so that large expanses of glass could be oriented primarily in a direction where the sun is

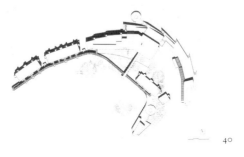

39 Rough Creek Lodge and Conference Center—model
40 Rough Creek Lodge and Conference Center—plan

relatively benign. Most of the outdoor terrace space is located on the north side to catch the views but also so as to be in shade for much of the day, sheltered by the buildings from the hot Texas sun. Only a small lower-level winter patio and upper-level porch are located on the south side of the building, protected from winter winds from the north and open to deep penetration by the low winter sun. A pool and fitness center is slid around the ridge, bringing it out of view from any of the other buildings and giving it full sun exposure from all directions.

All private rooms are given the north distant view with circulation placed on the south side. All major function spaces—living room, dining room, and meeting rooms—are likewise oriented north to ideal view and light. Auto access and circulation are tucked behind the buildings at the top of the ridge, separated from the lodge by a winding 650-foot-long stone wall. The wall is penetrated occasionally for gateways into the complex. It moves in and out between the existing oak trees, sometimes placing them as focal features in the patios along the circulation route.

Building massing, shape, and even function locations are driven substantially by the landscape and the geography of the site. The gentle curve of the buildings as they wrap around the ridge comes directly from the contours of the land. Even the shape of the building section in the main lodge space is a reflection of the soft rounded forms of the hills surrounding the site. The general zoning and location of functions have to do with concealing access and parking, optimizing views and sun orientation, creating outdoor spaces with maximum climatic advantage, and using the trees as space-definers. Site and buildings merge to form a continuous mass made of natural and man-made forms inextricably linked into a symbiotic whole.

The building's materials further tie it to the land. Limestone walls reflect the local geology which consists of deep chalky limestone strata. (A nearby town is named Chalk Mountain.) Bleached cedar lap siding, fair-faced concrete, and weathered standing seam metal roofs are compatible with the soft grays and tans of the landscape during the long Texas summer. Materials are assembled in a logical, tectonic manner. The stone piers of the main lodge building clearly indicate their construction—three tiers of stone formwork filled with three pours of concrete. A concrete shelf caps each pour and articulates the structural role of the pier as lateral bracing (larger at the bottom and smaller at the top) for the bowed roof trusses.

Interiors of the major spaces continue a connection with the character of the landscape. Major rooms are tall, open, and filled with light. The curved form of the main lodge is exposed inside in the form of laminated bow-string and king-post trusses with timber decking between. Columns, king posts, curved beams, and struts are all constructed of layered materials with spacing blocks, exposed connectors, and steel rods detailed to reveal the "craft" of their construction. Native interior materials—pine structural members, limestone walls,

41 Rough Creek Lodge and Conference Center—terrace

and mesquite floors—continue a harmonious dialogue with the site. Living room and dining room functions spill out onto broad terraces increasing the sense of wide open spaces in the interior. All rooms, both public and private, have some sort of outdoor space—deck, porch, terrace, or patio—associated with them.

The social values as well as the physical character of this memorable place are embedded in both interior and exterior spaces. The pace of life is slowed way down here. Paths meander and get easily sidetracked. All routes are enriched with places to stop and talk or look. Patios, terraces, porches, and verandas with comfortable furniture and a sense of ease encourage distraction. This is overtly a rural place, absent the intensity of a city.

The people in the nearby rural communities of Chalk Mountain, Glen Rose, and Hico are warm and friendly, welcoming to strangers. The buildings here are intended to project those same values. Spaces are relaxed and informal. Colors and materials exude a warmth and simplicity. The idea is to communicate a feeling of people's lives here without wandering into the realm of kitsch or nostalgia.

How is living fundamentally different day-to-day in a place where people drive their pick-ups 40 mph in a 70 mph speed zone and pull over to the shoulder politely when city folks roar by at a faster pace? How are people affected by living their whole lives under the magnificent dome of that big Texas sky with its myriad constellations at night and ear-splitting lightning and thunderstorms? How is our sense of ourselves as human beings altered by living in an environment where our impact is tiny and the forces of nature are huge? The placement of buildings, their orientations to views, the sense of isolation of almost every room in the complex, the intentionally inconvenient splitting of functional parts, all are intended to take visitors far away from their everyday environments and root them firmly here in North Texas.

At the time of its completion, *Architecture* magazine referred to Rough Creek Lodge as "an unpretentious place, inspired by the gentle hills and meadows that surround it, by native materials, and the vernacular of Central Texas." They further noted, "Rough Creek's regionalism is not a pastiche—but a harmonious mix of modern and longstanding forms and materials, applied with a consistently straightforward sensibility." The lodge buildings extend the strong and lasting sense of place inherent in the site and region by drawing inspiration from natural and geographical conditions as well as the everyday life of the place. Qualities of the local horizon, land contours, vegetation, light, textures, materials, and color all serve to inform the buildings. In the end, the project is about making the landscape and the sense of place stronger because the building is there—making the shape of the land, the ledge, the oak trees, and the big sky seem even more intense than they were before.

30 / ROUGH CREEK LODGE AND CONFERENCE CENTER

A

B

A. Lodge components wrap around a ridge with commanding views of a lake in the foreground and distant shapely peaks beyond. **B.** Breaks in the stone wall frame views to the north as well as providing access points for guest rooms. **C.** Nestled in and around a grove of live oak trees, the center is composed of a series of buildings of varied sizes and shapes drawn together by a 650-foot-long stone wall.

C

D

D. South façade features a deep shady porch and a winter terrace protected from strong north winds.

E.

F.

E. Smaller structures like the gun room and fire pit are located at the edge of the complex to create a transition to the natural landscape. **F.** Broad north-facing terrace connects the lodge's most public spaces with views and landscape. **G.** Framing of buildings is logical and tectonic emphasizing methods of assembly and differing structural role of wood, steel, stone, and concrete. **H.** Interior public spaces match the monumental scale of the landscape outside. **I.** Generous amounts of glass, carefully oriented and thoroughly shaded, allow connections between indoors and outdoors without excessive heat gains. **J.** Porches, terraces, and patios occupy almost as much footprint as the buildings, providing comfortable, habitable spaces outdoors.

G

H

I

J

L

K. As carefully defined and articulated as the building's interiors, outdoor rooms are an integral part of accommodating the lodge's program. L. Private outdoor spaces match the attention given to larger decks and verandas with light, delicate materials and details. M. Rough Creek Lodge creates a powerful link to both its site and its geographical region through its natural responsiveness to circumstance and exigency.

K M

02/

A BROADER VIEW OF SUSTAINABILITY

A BROADER VIEW OF SUSTAINABILITY

In the late 1960s, Paul Rudolph, who was, at the time, one of the most respected American architects of his generation and who was dean of the School of Architecture at Yale University, wrote the following: "In architecture all problems can never be solved.... Indeed it is a characteristic of the 20th century that architects are highly selective in determining which problems they want to solve."9 He posited that great architects make wonderful buildings only because they ignore many aspects of a building. He noted, "If we solved more problems, our buildings would be far less potent."10

Rudolph's words epitomize one of the most powerful aspects and one of the most disturbing problems of 20th-century thinking. Rudoph overtly favors narrowing the broad scope of an architectural problem in order to make it clearer and more manageable. He advocates concentration in order to obtain power and potency. He favors exclusion as a means to achieve intensity.

This attitude in architecture is consistent with 20th-century notions in many disciplines where reductivism, oversimplification, and single-mindedness have led to negative consequences which we must now address in crisis proportions. In science, in politics, in education, even in art, we have sought clear, singular models, resisting complications as "static" in the system. In our frustration with the inability of those models to produce success, we have jumped from one model to another sequentially, searching for that one satisfying, clear, distilled direction.

In architecture, this reductivist attitude has produced a whole series of movements—Modernism, Post-Modernism, Late-Modernism, Neo-Modernism, Metabolism, Historicism, Brutalism, Deconstructivism, etc.—each concentrating on one or a few aspects of building, but none taking a really broad, synoptic view. The current interest in sustainability could be seen as just another one of those movements—something that will be "hot" for a while but will soon go out of fashion as some other set of interests supplants it.

But perhaps the destiny of this movement is different. Perhaps the 21st century represents a more broad-minded kind of thinking, that is not content to ignore many aspects of a building in order to concentrate on a narrower band of interests. Perhaps sustainability is, in fact, precisely about seeing architecture as a broad, multifaceted endeavor which is greatly diminished when its scope is constricted.

I was very encouraged recently by reading an article written by two British researchers in an issue of the *Journal of Architectural Education*, edited by Kenneth Frampton and Steven Moore. Entitled "Reinterpreting Sustainability Architecture," it does an impressive job of describing the full breadth of what sustainability currently means to various constituencies. Six different "logics" of sustainable architecture are described, covering issues as diverse as social equity, community participation, quality of life, health factors, expression of local cultures, regionalism, ecological consciousness, biodiversity, compact cities, and energy conservation, among others. The authors suggest abandoning "the search for a true or incontestable definition of sustainable buildings" in favor of just embracing its whole messy range of concerns under the broad umbrella of sustainability. They suggest that the pursuit of "consensus that has hitherto characterized sustainable design and policy making should be translated into the search for an enlarged context in which a more heterogeneous coalition of practices can be developed."[11]

It is actually this inclusive attitude toward diverse design concerns that makes the notion of sustainability particularly appealing to me. Here is a way to look at architecture in the very largest frame of reference. How does it perform socially, functionally, aesthetically, and technically? How does it perform over a multigenerational time frame? How does it embody timeless human values like responsibility, intelligence, perceptiveness, creativity, and ingenuity?

For me, the beauty and the challenge of architecture is that, in the end, it is about life—all of life. Sustainability seems to accept that notion. Architects must be concerned with how many watts of electricity per square foot the building is consuming, but they must also be concerned about whether the lighting produced by that electricity is creating an atmosphere supportive of the functions that occur within the building. Designers must deal with conserving natural resources in the construction process, but they must also be sure they are producing a building which will have a long, productive functional life—which will not become obsolete and require premature replacement, negating the original efforts to conserve. Architects must avail themselves of opportunities to recycle and provide the option for future recycling in the choice of building materials, but they must also pay attention to whether the materials they select are contributing to promoting employment stability, developing craft and local culture, and engendering a sense of identity and community pride.

The broadest agenda of sustainability can never be fully satisfied or even fully optimized. Its range is so inclusive that some issues will inevitably even conflict with others. In the recent exhibition "Ten Shades of Green," sponsored by the Architectural League of New York, ten exemplary buildings were chosen from around the world to represent the best of green architecture. Though each embodied great strengths in terms of sustainability, there was also room for criticism in every case. In sustainable design, perfection is not a realistic possibility. Balance, appropriateness, ingenuity, synergy, inventiveness, intelligence, and sensitivity are better criteria for evaluation.

In our own work over the last 20 years these have been our watchwords. We have always sought to make buildings which are appropriate to their place, their purpose, and to the people who use them. We try to make buildings that are poignantly expressive of the local values and

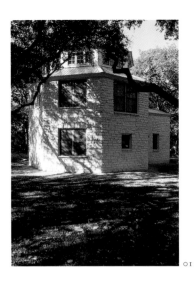

01 Private Library, Westlake Hills, Texas

culture of the populace they serve. To do this in a sophisticated way requires a strong degree of community involvement and community participation which we sincerely enjoy.

We pride ourselves on our ability to site buildings well so that they make a positive impact on the urban condition or on the landscape around them. We worry a lot about the building's interaction with the sun, the wind, and the local ecology. We are trying, in every project, to create a symbiotic relationship between indoors and outdoors—between man-made and natural.

We love building materials and we have always researched them thoroughly—where they come from, how they are made, how they might be used in a particularly evocative and sensible way. We have tried desperately, no matter what the budget, to build well—to make places that will last a long time and have real durability—technically, functionally, and aesthetically.

We did all of these things long before the term sustainability came into common usage. In the very first commission I ever did on my own—a tiny private library in West Lake Hills, Texas—issues now associated with sustainability were already of concern to us. The library was kept to a minimum footprint and was organized vertically in order to nestle within the canopy of a grove of large live oak trees. It tapered at the top so as to become the inverse of the tree's boughs and avoid any substantial pruning. The natural environment was deemed primary and the little building a tenant that avoided intrusiveness.

Daylighting was important to us, though we did not call it that. We simply preferred natural light to artificial light as a resource for reading as well as other activities. We tried to make a relatively glare-free reading environment, learning explicitly from lessons of Alvar Aalto's extraordinary Viipuri library of the late 1920s. We brought light from high in the room and from all sides evenly to minimize reflections and to allow a reader to shift positions in the room or chair without losing optimal light.

Minimizing resource consumption was also a high priority. Because the site was shady and located on a knoll that caught every available breeze, we tried to make a building without a mechanical cooling system. We placed well-shaded operable windows on all sides of the polygonal room to catch every wisp of wind. We also put operable windows high in the lantern so that as warm air would rise it could be exhausted, drawing in cooler air low from the shady understory of the grove. We even took the challenge of heating the building without a mechanical system, installing a recirculating fireplace with ducted distribution of warm air from a plenum around the firebox directed to other parts of the room through simple convection. In 1984 the little library won an energy conservation award where one was required to submit utility bills for heating and cooling. We had nothing to submit.

Materials were earth-friendly and very local. The wood for

02 Private Library—interior

framing as well as even finish carpentry for bookshelves and balustrades was farmed East Texas Pine. Limestone for both interior and exterior walls was quarried about 20 miles from the site. Stone flooring was a refuse material from the quarry created when large rough-hewn blocks were first sawn to dimensional units. We turned the sawn sides of the refuse pieces up for the floor, placing the rough side down into the grout. Fieldstones for the fireplace were gathered from the site.

Neither we nor the client was trying to make any grand statement. We were just building in a way that seemed sensible, economical, and sincerely respectful of the land we were inhabiting. Twenty years later, the client is still enjoying the little library immensely. Though we did make provisos for the addition of mechanical systems, they have never installed them. They live with more temperature variation than most people in a coddled era would like, but it is impressively temperate in the space even on a hot summer day or during a rare cold winter spell when they are obliged to keep firewood picked up from the site and the fireplace stoked.

As this broad definition of sustainability has emerged over the last several years, we have embraced it and have benefited enormously from the dialogue it has provoked. As the two following projects demonstrate, the issues become far more complex and difficult at the scale of large buildings and when dealing with institutional clients than they were with the little library. But the passion for economy, simplicity, straightforwardness, and a genuine respect for nature and natural forces remains the same.

Robert E. Johnson State Office Building

Located on a very visible site a block away from Austin's State Capitol Building, this prominent 320,000-square-foot civic structure was designed to be home for eight of Texas' legislative service agencies. From the very beginning, it was sponsored by the Texas State Energy Conservation Office to become a model of sustainability for future state office buildings. A broad range of consultants, including the Center for Maximum Potential Building Systems of Austin and Berkebile Nelson Immerrschuh McRowell of Kansas City, were involved in research and systems development.

From the earliest site-planning stages, a commitment was made to filter every design decision through a sieve of sustainability issues. Though an earlier master planning effort had suggested a squarish office building set in the midst of a ceremonial lawn like most of the other recent state buildings in the area, an immediate effort was undertaken to get permission to alter the master plan both to produce a thinner, more exterior-focused building, and to create better scaled, more usable outdoor spaces.

The approach we suggested was to break the building into two thin bars—one creating a strong, formal edge to Congress Avenue (the

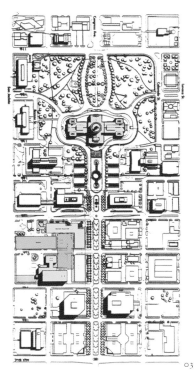

main spine of downtown Austin) and the other defining a somewhat more informal boundary for Fifteenth Street (a major east/west arterial). The two bars, along with an earlier-phase garage, would then define an intimate north-facing court which could form an internal green space for the complex. The conference center, a part of the original program meant to serve not only this building but adjacent ones as well, was pulled free of the two office blocks and allowed to occupy the east edge of the court. This accomplished three goals. It shrouded the harsh edge of the parking garage; it provided a smaller-scaled element to create a more intimate feeling in the court; and it created constant activity in the court as people came and went to meetings and as they took breaks on the conference center porch which faced the court. The shapely conference center pavilion was angled slightly so as to link the new court with a pleasant green space dominated by two huge live oak trees across the street.

On the Fifteenth Street side of the building there were several large oak and pecan trees that meandered onto the site from the street space. The office bar there was carved away to accommodate the canopy of the trees as well as to provide additional light wells intermittently for deeper daylight penetration in the long south façade. This also helped break down the scale of the massing on the less ceremonial Fifteenth Street frontage. At two of the light wells, the ground floor is hollowed out straight through the building for a breezeway focused on a particularly fine live oak in one instance and for a sallyport in the other.

The result of all these site-planning moves that optimize usable outdoor space and create features out of existing vegetation is to make the whole environment feel softer, greener, and more natural than it would have otherwise. Occupants looking out on the court enjoy quiet natural vistas. Windows opening onto light wells filled with mature trees create an entirely different work experience than if they looked out on busy Fifteenth Street directly. One goal of sustainability is to keep human beings more connected to nature and natural forces. Even on this very urban site, small urban design moves went a long way toward accomplishing that goal.

Site-planning and ground level treatment also help promote amenable pedestrian movement in the district. The breezeway and courtyard serve as convenient and pleasant cut-throughs creating shady paths from the parking garage to the Capitol Building or from the conference center toward nearby office buildings. Covered arcades line two edges of the building where pedestrian movement is frequent, providing pleasant sheltered pathways for exterior circulation. Arcades, the breezeway, the court, and other terraces are all linked and carefully related to entry points, lobbies, and even staff lounges to invite easy everyday movement and interaction between indoors and outdoors.

The general massing of the buildings was also fundamental in helping to accomplish another important sustainability goal—natural

03 Robert E. Johnson Building, Austin, Texas—site plan
04 Robert E. Johnson Building—breezeway

05 Robert E. Johnson Building—courtyard
06 Robert E. Johnson Building—ground floor plan

daylighting of as many office spaces as possible. The thinner section of the building bars and the light wells on the south face made natural light much more available than the massing of the master plan scheme. Large apertures on this expanded building face were carefully articulated in response both to exterior orientation and to their role in lighting rooms within. Windows rise from desk height to 12 feet above the floor. There is a thick horizontal mullion at the eight-foot six-inch level which contains perforated blinds controlled by the users. Below the eight-foot six-inch line the material is heavily treated low-emissivity and ultraviolet protective glass on the south and west faces to maximize daylighting without getting excessive heat gain. On the long south face there is also a light shelf at the eight-foot six-inch bar with a specular metal surface on its reflector side which catches light from the upper glass pane and bounces it deep into the building. Though still a low-emissivity glass, this upper pane is rated for a bit higher light penetration than the lower pane. The flow of natural light through the building is enhanced by clerestory windows above seven feet on interior partitions parallel to outside walls. These allow for privacy at eye level, but provide deep light penetration through the building above.

Fundamental to the success of the daylighting strategy in the building were the tall ceiling heights at the building's perimeter, essential for bringing light in high at the edge and distributing it deep into the building's center. The height is accomplished, in part, by close coordination of mechanical systems to virtually eliminate continuous

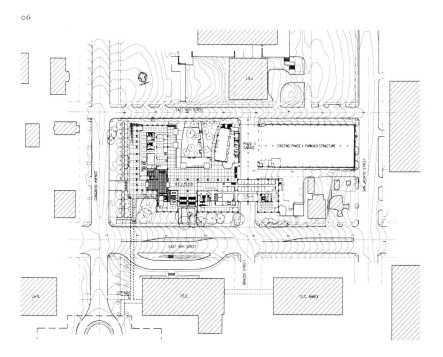

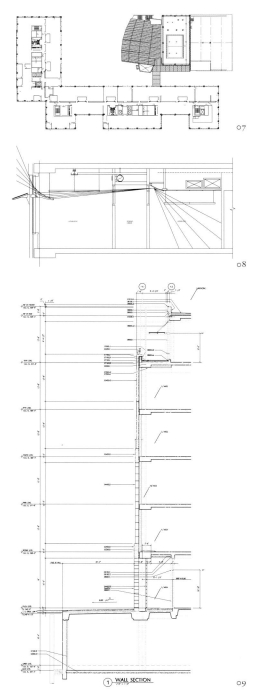

07 Robert E. Johnson Building—upper floor plan
08 Robert E. Johnson Building—office sections showing light shelf
09 Robert E. Johnson Building—wall section

plenums or hung ceilings at the edges of the building, and, in part, by utilizing a thickened flat slab structural system which eliminated the need for deep perimeter beams. It is in this coordination between structural, mechanical, electrical, and environmental characteristics of a building that fundamental sustainability advances are made.

At every step, technical systems of the building were pressed for greater efficiency and reduced resource consumption. Electrical loads, especially for lighting, were cut wherever possible. Coupled with the daylighting, dimmable ballast fixtures and automatic sensor switching devices drastically reduced electrical requirements. Lessening heat generated by artificial lighting along with careful fixture and equipment selections allowed reduction in required cooling tonnage. Performance monitoring systems and a thorough commissioning process further ensured efficiency of operation.

Careful and deliberate use of building materials was also an overriding theme used to produce both high standards of building performance as well as strong architectural character. Inspired by the powerful materiality of the 1888 granite State Capitol Building down the street, the Robert E. Johnston (REJ) building was conceived as an architecture based equally in material technology and appropriate expression of purpose and place. The building's materials became the essential common denominator addressing a broad range of issues from contextuality to sustainability.

Design decisions at every level—from the frame, to the skin, to interior finishes, to furnishings—were shaped by a constant consciousness of how appropriate materials usage could guide and inspire the design process. A robust concrete structural system composed of rectangular columns and flat slabs thickened at concentrations of structural loads was conceived, not only to be economical and long-lived, but also to give the building a feeling of heft and permanence for its public purpose. Concrete, as opposed to steel, offered the opportunity to expose the structural system inside and reveal the strong, durable nature of the building even in office environments which are so often dominated by a feeling of transience and impermanence. Exposing the concrete frame also accomplished the sustainability goal of reducing redundant systems with their wasteful duplication of materials. Avoiding hung ceilings wherever possible resulted in a reduction of initial resource consumption as well as elimination of one of the elements of building assembly which deteriorates most rapidly. The concrete flat slabs also facilitated the accomplishment of other sustainability goals, as mentioned earlier, by reducing structural depths at the building edges allowing greater daylighting opportunities.

The concrete itself was made more "earth-friendly" by replacing part of the cement (35 percent in some portions of the building, 28 percent in others) with fly ash. A finely divided inorganic residue which is a waste byproduct of coal combustion, fly ash offers the oppor-

10   Robert E. Johnson Building—granite blocks for load-bearing wall
11   Robert E. Johnson Building—building skin
12   Robert E. Johnson Building—courtyard

tunity to find a very useful and economical purpose for a material which otherwise ends up in unsightly ash dams dominating the industrial landscape. (In 1998 the U.S. produced 44.9 million tons of fly ash.) Replacing cement with fly ash not only helps alleviate a serious waste disposal problem, but also reduces consumption of cement and all the energy resources and toxicity required to extract and process it.

The building's skin, like its frame, grows out of a careful investigation of the rich potential contribution of materials to both building character and performance. Granite was selected as the predominant exterior material for both its toughness and its visual qualities. The pink native Texas stone ties the REJ building closely to the adjacent State Capitol complex and also appropriately bestows the attributes of dignity and durability on a structure housing government agencies.

The granite was quarried about 50 miles from the site, in Granite Shoals, Texas. The actual configuration of the stone is very unusual for a contemporary building. The granite is load-bearing, carrying its own weight rather than being hung from the concrete frame. Stone thickness tapers from 14 inches on the ground floor to eight inches at its highest point. This reinvigoration of a time-tested construction method was employed both to exploit the inherent strength, mass, and weight of the granite and to reduce dependency on vulnerable metal hangers concealed in moisture-laden cavity spaces. Stone finishes are either cleft or sawn, techniques chosen over more energy-consuming processes like polishing or flame-finishing.

The granite is stacked in vertical piers with spanning elements made of precast concrete—a more contemporary local material reliant on abundant Central Texas aggregate supplies. An aluminum curtain-wall system was selected after careful analysis of energy performance, location of source materials, durability, and available finish properties. Spandrels in the curtainwall are zinc, chosen for its longevity and non-toxicity as well as for its color and texture, which complements the gray-flecked granite.

Inside the REJ building, a similar set of concerns prevailed for the generative role of materials in creating a healthy, supportive, and responsible environment. Wood, which is employed in both the ceilings and walls of public spaces for its richness and warmth, was specified to be from sustainably managed sources that are "smart wood certified." Wood products throughout the building, including medium density fiberboard (MDF) employed in extensive cabinet work, use no urea formaldehyde. Metal studs in partitions have 65 percent recycled content. Gypsum drywall has 100 percent recycled paper on face, sides, and back, and partial recycled gypsum in the core. All paints and other finish coatings are water-based with low- or no-volatile organic compounds.

Colored cementitious toppings are used for hard surface flooring. The choice of carpet stemmed from the manufacturers' commitment to sustainable manufacturing processes. Specific product selection was

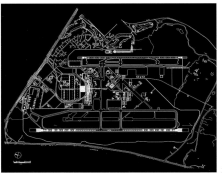

13 Austin Bergstrom International Airport, Austin, Texas—massing model
14 Austin Bergstrom International Airport—site plan

based on high-wear performance, recycled material content, and ability to minimize use of adhesives in installation.

The design team devised a matrix to evaluate the appropriateness of various materials under consideration. Factors rated included where the material was extracted and how it would be transported to the site, toxicity of the material both during manufacture and after installation, recycled content as well as recyclability, and life-cycle costs. The same matrix was used by the team who selected furnishings, resulting in the use of almost 13,000 yards of fabric made from recycled polyester and plastic bottles for workstations as well as a wide range of other new products drawn from manufacturers employing manufacturing protocols that drastically reduce pollutants.

This kind of thorough, "all-issues" approach to sustainability requires a strong team effort including the client, architects, engineers, and consultants. It requires injecting sustainability considerations into virtually every design decision as well as into technical considerations. It necessitates a great deal more communication and coordination among team members than in a more normal design process.

Austin Bergstrom International Airport

This new passenger terminal for Austin is a 25-gate "start from scratch" facility which has 680,000 square feet of space on three levels and cost $140 million. It was the first major passenger facility to be constructed in the U.S. after Denver International Airport opened in 1994. It took two and a half years to design the project and three years to build, and was completed in July 1998.

The following are seven sets of sustainability issues that we explicitly addressed. These are not comprehensive by any means, but they will give a flavor of the way in which this broad, inclusive notion of sustainability might impact design. The first set of issues had to do with treatment of the site.

The site is a former U.S. Air Force base. The new airport was located here in order to reuse a very high quality 12,250-foot existing runway, so that even the decision about locating the airport was an act of recycling. Every effort was made to reuse as many buildings on the Air Force base as possible. The main administrative building was converted to a hotel and other ancillary buildings were used for Aviation Department offices. Homes for military personnel on the base were moved to become city-owned low-income housing. Mature trees were carefully surveyed and retained where possible, but over 30 large live oak trees which could not be kept were successfully moved and replanted on-site.

All demolished concrete on the site was recycled as sub-base material on other airport paving projects. All demolished asphalt was recycled as fill. All excavation materials were reused for other projects on the site. Insofar as possible, the invested resources on the site were

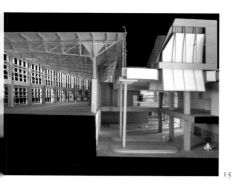

15 Austin Bergstrom International Airport—sectional model

respected and utilized to the fullest extent. Two basic tenets of sustainability—reuse and recycle—led decisions about the site throughout the project.

A second set of sustainability issues dealt with shaping the footprint and massing of the building on the site. Initial studies for the location and shape of the building were dominated by a pair of important environmental concerns. First, we were intent on keeping the amount of site regrading to an absolute minimum. We also wanted to avoid dramatic changes in patterns of site runoff and rainwater evacuation. Though this was a relatively flat site, there was still a significant fall across the enormous footprint necessary for the terminal apron and runways. The long, thin shape of the terminal and its orientation were conceived to cut across as few contours as possible requiring dramatically less cut and fill than the initial master plan had called for.

The natural fall of the site was used to our advantage by making the long, thin terminal building a retaining wall and creating a section which nestled into the slope. The airplane apron could then be higher than the baggage claim area, providing a very efficient path for baggage conveyor belts.

Another significant environmental factor shaping the footprint of the building was sun orientation. We knew from the beginning that daylighting would be a very important factor for us in achieving both energy efficiency and the appropriate "feeling" and "spirit" for the terminal. The building was, therefore, stretched very long on the north/south faces and very short on the east/west faces. In our climate, north light is gentle, relatively benign and very useful. South light, if properly shaded, can be beneficial for maximizing light in the winter and keeping out direct sun in the summer. East and west sun are more difficult to control, and west sun especially contributes problematic heat gain in late afternoon in the summer when it is most harmful. So, in the end, the building is almost ten times as long as it is wide in its thickest part. It is an almost perfect sun and light catcher for our climate.

A third critical set of sustainability issues had to do specifically with heating, cooling, and otherwise providing technical systems for the terminal. Natural light is a very important theme for the building and an essential element in conserving energy. In this internally loaded building type, where heat is normally generated primarily by artificial lighting, machines, and people rather than heat gain from the exterior skin, we sought to minimize heat produced by light fixtures. By maximizing the benefit from cooler natural light, we were able to significantly reduce reliance on artificial light.

Utilizing readily available insulated glass technology we could employ the natural light without excessive heat gain. We have over 102,000 square feet of glass of three different types. One has a ceramic frit. The others are varieties of low-emissivity glass. Windows on the south have six-foot-deep horizontal sun shades as well as a concentration

16 Austin Bergstrom International Airport—entry

of the higher-performance low-emissivity glass. Windows on the north side need no sun shades and can even have a somewhat less protective glass. The various glass types and sun shades make a subtle, but complex pattern on the south building façade facing the airfield, which contributes visual interest as well as energy performance.

Many other energy-saving strategies are employed in the building: a reduction of evening ambient light levels with concentration on task lighting in hold rooms and eating areas; sophisticated automated controls including occupancy sensors, photocells, and timers on both lighting and heating/cooling systems; a large ice-storage capability which allows off-peak energy usage; and high-efficiency motors with variable frequency drives in mechanical systems, to name a few. The building was conceived—from its siting to its massing to its skin to its technological systems—as a great efficient machine which first relies on full utilization of natural forces and then supplements these with the most resource-conscious mechanical means possible.

But it was certainly not the case that the airport building was viewed simply, or even largely, as a technological machine. Our primary goal was to make it a very personal, user-friendly place which would be embraced by the citizens of Austin as an important public place and a powerful symbol of the city.

In this regard, a fourth set of sustainability issues in the project had to do with design process and incorporating a broad range of input and involvement. Our client was, in fact, the citizens of Austin who voted the bond money for construction. We worked very closely, of course, with the mayor and City Council, who, as elected officials, represented the citizenry. In addition, a committee of 22 people was appointed to work with us in a much more detailed way. This committee worked very hard over a period of almost two years. As representatives of many diverse constituencies in the city, they helped us understand various points of view. In addition to working with this special committee, we also made dozens of presentations to a range of other groups and committees in the city, broadening community participation even further.

But the client for the airport was not just the public and their representatives. The building had to respond to very specific needs of the airlines, the rental car companies, concessionaires, and various other contractors and vendors who operate out of the terminal. Literally hundreds of hours were spent in meetings with the airlines and other private business ventures trying to understand their specific needs in this building. Going through such an elaborate and complicated input process is essential in order to fully embrace a fifth set of sustainability issues that have to do with building function.

Virtually everyone we spoke with was unhappy with most other airports they had visited or worked in. They strongly emphasized the importance of convenience for passengers. Austin is an origin and destination airport—not one where people are changing planes, but a

17 Austin Bergstrom International Airport—"Marketplace"

place where more than 95 percent of passengers are going from ground transportation to the gate or vice-versa. This means the challenge was to create the very shortest and most direct path possible between the curb and the gate. The basic diagram of the building optimizes that path.

Two entrances at the curb place passengers at the most central access point for all the gates. The building is divided into two distinct parts—landside and airside. As one enters the landside, ticket counters are immediately visible on either end. Baggage claim is in the center in an open well. Straight ahead are security checkpoints, and just beyond that, concessions and the gates.

The nine gates in the crescent-shaped portion of the terminal are used for commuter flights—short hops, often to nearby Houston or Dallas, which depart every hour. The distance from the curb to these gates is as efficient as possible. Many business people in Austin take these flights two or three times a week for day trips and such ease and efficiency is greatly appreciated. Concessions are located immediately adjacent to the hold rooms in the great central space we call the "Marketplace." The process of going through security, checking in at the gate, grabbing a bite to eat or a magazine while waiting to board, and then loading the plane is greatly facilitated by this easy, convenient orientation of parts. The atmosphere in the Marketplace is lively and urbane. There is café seating dispersed casually throughout the space. A live music stage is in the center of the crescent where Austin's famous music scene is on display in an easy, unpretentious way. There is a good feeling here—friendly, warm, open, and inviting.

Another prominent functional concern in the design of the airport was wayfinding. We tried hard to make the building very transparent and legible to users. We wanted to create a place where people could see or intuitively sense where they needed to go via cues given by the building. From the entry points we tried to make as many airport functions as possible visible immediately so as to aid in orientation.

Ticketing lobbies are clearly available to one side. Security checkpoints are straight ahead. Concessions and gates are visible just beyond that. Even baggage claim can be easily located down below—not in the kind of low-ceilinged basement common in many terminals, but in a bright, open room which contributes to the liveliness and "action" of the larger building.

As passengers move through the terminal, the notion of easy wayfinding continues. The crescent shape helps to locate them in the large central room. Even as passengers first exit the airplanes, they know where they are in relation to the center and the entry/exit points, just by the shape of the space. Wayfinding is, of course, additionally supplemented by clear, well-located signage.

A sixth set of sustainability issues which helped formulate the airport's design had to do with materials selection. At every stage an effort was made to employ materials which had low levels of embodied

18 Austin Bergstrom International Airport—steel frame
19 Austin Bergstrom International Airport—ceiling
20 Austin Bergstrom International Airport—detail

energy, came from renewable sources, minimized negative impact on air quality both in the manufacturing process and as installed, and would last a long time with minimum reinvestment in maintenance and replacement. These were tough issues to deal with in an airport where use is both demanding and constant.

Steel for the project, which is the predominant structural material, is 95 percent recycled. Concrete, which is used primarily for piers providing lateral bracing, employs fly ash to replace part of the cement. The primary finish material both inside and out on the landside portion of the building is a local Texas granite quarried about 50 miles from the site (similar to that used at the REJ Building, but a cooler, grayer color). Its toughness in an environment of carts, baggage, and jet fuel exhaust is crucial. It is used in thick slabs to increase durability and longevity. Corners are made of solid L-shaped pieces for protection from impact. Because the granite is local, embodied energy from transportation is minimal. We were able to work very closely with the quarry to employ the material in a way that minimized waste in the production process and gave us excellent color and texture consistency.

Inside the terminal, every effort was made to minimize redundant systems with their wasteful duplication of materials. Avoiding hung ceilings wherever possible resulted in a reduction of initial resource consumption as well as elimination of one of the elements of building assembly which deteriorates most rapidly. Steel decking and framing becomes the finish material throughout much of the public space of the building.

Where interior finish materials are required, they are chosen to complement the steel frame and granite piers and walls. Wood is used extensively, but always high on walls or ceilings so as not to be vulnerable to impact damage. All wood products were supplied from controlled growth forests. Much of the warm, wood-feel of the building actually comes from MDF (medium density fiberboard) which is made of waste material from wood processing. It is used frankly with a clear finish which exposes its character as a wood material, but also a modern industrial product.

Another common wall material in public spaces is sisal—a natural fiber which is tough like rope. It can easily take the abuse required, but is also warm and rich in its color and texture. Seams, which are the least durable part of a sisal assembly, are avoided by placing sisal panels in an aluminum framework. The metal angles serve to keep carts and bags from banging against the wall as well as to protect vulnerable joints. The idea, well-illustrated in this assembly, is to use natural materials with low embodied energy, detailed in such a way as to last for many years and not require wasteful replacement.

For areas where eventual replacement is almost inevitable, our emphasis was on recycling. Toilet partitions and benches, for example, are made of recycled materials. Carpet, which is only used in low-traffic

21 Austin Bergstrom International Airport—Texas rivers terrazzo mural
22 Austin Bergstrom International Airport—street names terrazzo mural

areas, is all recyclable. In these and other manufactured products there was great concern with chemical content. We as architects avoided urea formaldehyde, for example, in the MDF. We used paints and other coatings which were water based and have low or no volatile organic compounds. Throughout material selection we kept a keen eye on where materials came from, how they were manufactured, how long their life expectancy might be, and where they might be reincarnated into future useful products.

The seventh and last set of sustainability issues we addressed had to do with the building embodying a sense of place. I was delighted that one of the six "Competing Logics" of Sustainable Architecture described in the *Journal of Architectural Education* article mentioned earlier focused on buildings responding to "regional context" and adapting themselves to "local physical and cultural characteristics."[12] In order for a building to be truly sustainable, it must become an integral part of its physical and cultural context. It must embody a tangible sense of locality and place.

If we believe in the conservation of the richness and diversity of life on this planet, that belief should include protecting human cultural diversity. Any model of ecological sustainability must incorporate a retention of meaningful local cultures which enrich and sustain the long-term health and breadth of perspective on the planet. We are, in fact, very fond of the local physical and cultural context in Austin and enjoy very much the opportunity to try to infuse some of its vitality into architecture. The airport was an extraordinary opportunity in this regard. This is the gateway to our city and the one single place where more citizens come and go every day than any other. If any building should embody the culture of Austin, this one should.

We as a design team explicitly tried to incorporate several cogent characteristics of the region in the design of the airport. The natural landscape in the central Texas Hill Country is very rich and particular. It is an unusual amalgam of toughness, ruggedness, and strength alongside beauty, frailty, and vulnerability. We Austinites dearly love our landscape in Central Texas, and we like to spend a lot of time outdoors inhabiting it. Our creeks, rivers, and streams especially often take on mythic proportions becoming resources for recreation and social life. We have focused much of our identity as a city on outdoor places like Barton Springs Pool and Town Lake Park where the natural environment has spawned an open, friendly, congenial way of life. We are a real outdoor sports and recreation city, but with an emphasis on participating sports rather than spectator sports. We are one of the few cities our size in the U.S. with no big-time professional sports teams, and our big local athletic hero is bicyclist Lance Armstrong.

We have a very diverse population ethnically and, mostly, we thrive on the differences in our backgrounds. We're a pretty tolerant city with a great appreciation for our various heritages and traditions.

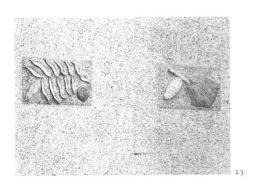

23 Austin Bergstrom International Airport—carved glyphs
24 Austin Bergstrom International Airport—stair rail detail

I think it is partly that same tolerance and appreciation that has generated a very lively and eclectic music and entertainment scene in Austin that ranges from country and western to blues and jazz to pop and rock. The Sixth Street Historic District downtown is our party room and the focus for a very vital and vibrant public life.

We are also the state capital, home to the state's flagship university and a long-time center for history and government as well as for research and education. In the last few decades we have extended those longstanding roles to become a center for high tech industry. Dell Computers began and is currently headquartered in Austin, along with hundreds of other software and hardware firms. But technology in Austin extends well beyond digital and electronic technology. As a city we depend on intellectual capital. We have a highly educated populace with a real bent toward logic, analysis, and creativity. These are all very challenging cultural characteristics to try to incorporate into an airport.

As architects we addressed Austin's love of nature mostly by creating a building that opens generously out to nature and the outdoors. As one observer was quoted when the building opened, "It feels like you are outdoors even when you are indoors." This is in stark contrast to many airports which seem to be an interminable maze of corridors and internal rooms. We even thought of the big trusses in the central space as tree-like with lots of complex pieces growing out of a central trunk and providing a canopy for the well-lit space below. We planned that big, almost outdoors space to house the kind of informal social scene we so appreciate in Austin, and that has really happened. There is a very "Austin" kind of atmosphere—great food provided by local eateries like Salt Lick Bar-B-Que, Matt's El Rancho Mexican Food, and Amy's Ice Cream in a comfortable café kind of setting. It is an unusually relaxing and stress-free airport environment which seems right for a city which prides itself in being "laid-back." This is emphasized again by the live-music stage at the center of the Marketplace.

Another tie to nature comes, of course, in the extensive use of the local granite. It delivers the sense of toughness and ruggedness in our landscape, especially when placed in contrast with sleeker, more refined materials. We commissioned a series of carved glyphs in the granite executed by local artisan Phillipe Kleinfelter who is a real expert in the carving traditions that are a part of our local Mexican heritage. The glyphs depict leaves and seed pods of local trees—the same series of local trees like live oak, pecan, and mesquite that were used to name the original east/west streets in the 1839 plan of Austin. The story of that original plan and its references to nature—not only in naming east/west streets for trees but also in naming north/south streets for the rivers of Texas—is told in a series of terrazzo murals on the floor in the baggage claim area.

Throughout the airport, artwork and craftsmanship are used to embody cultural values. Murals depict Enchanted Rock, a landmark

25 Austin Bergstrom International Airport—"Marketplace"
26 Austin Bergstrom International Airport—plans

granite formation made of the same stone as the building, or a lyrical series of scenes of a Hispanic family having a backyard picnic. An outdoor sculpture is based on seed pods of local flora. The grab-bar on stair rails is made of a hand-forged piece of steel made by local artisan Lars Stanley. It gives a softer, more humane touch where the body comes in contact with the building. The handrail in general provides just the kind of high-tech/high-touch combination Austin is known for.

The intellectual/analytical/high-tech ethos of Austin is present everywhere in the building. There is a clear tectonic nature of the structure which expresses logically how it is put together and how it works in a technical sense. It is also a precise structure, clean and exact. One observer described it to me as "mind candy" with lots of interesting pieces and details to be intrigued by.

In all of our work we are trying to approach architecture as a pluralistic discipline. We refuse to ignore many aspects of a building in order to make our buildings more potent as Paul Rudolf advocated. We revel in the various competing logics of design and enjoy very much grappling with the messy conflicts inherent in any full-bodied design process. We pride ourselves in making buildings that are responsive and particular to their circumstances and are not just signature pieces of their designers.

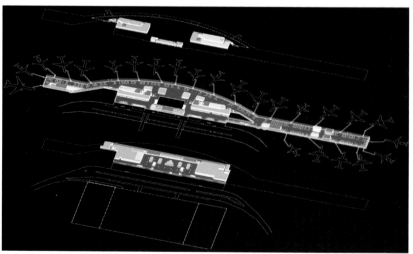

56 / ROBERT E. JOHNSON STATE OFFICE BUILDING

A. The building's large volume on a tight urban site is utilized to define a soft, green outdoor space as a focus for the complex.

B

C D

B. Street-defining face on Congress Avenue defers to the State Capitol Building a block away.
C. Locally quarried Texas Pink Granite in an unusual load-bearing pier configuration is used with precast concrete and zinc spandrels on exterior façades. D. Breezeway links new courtyard to a cluster of pre-existing live oak trees while also accommodating desire line for travel between the conference center and the Capitol building.

E. Ceilings are exposed to the underside of slabs through most of the building to facilitate deeper penetration of natural light. F. Even internal corridors are partially daylit, supplemented by complementary artificial light. G. Wood in lobby, breezeway, and arcade ceilings, specified to be from sustainably managed sources, adds richness and warmth. H. Light shelves on the south face of the building bounce daylight deep into the building while also protecting peripheral spaces from glare.

E F H

G

62 AUSTIN-BERGSTROM INTERNATIONAL AIRPORT

A. The high-ceilinged, daylit "Marketplace" at the center of the terminal has the kind of lively, informal-but-urbane atmosphere that its city is known for.

B

C

B. Wayfinding is transparent and intuitive with most airport functions visible as visitors enter the terminal. C. The baggage claim area is a bright open room that contributes to the sense of liveliness and "action" in the terminal as a whole. D. North light on the entry side of the terminal floods both ticketing lobbies and car rental areas below with resource-efficient diffuse lighting. E. South light, controlled by horizontal sun shades and treated glass, enters the terminal both at the level of passenger hold rooms and, through large clerestory windows, high into the "Marketplace."

F. Generous light wells allow daylighting of important functional outdoor spaces as well as building interiors. G. The sweeping curve of the primary indoor space helps orient visitors arriving at gates at the center of the building where they can exit the secure part of the airport. H. Wasteful redundancy of building systems is avoided by allowing steel framing and decking to be finish materials through much of the building. I. The simple, straightforward volume of the building and its effectiveness as a great "catcher" of natural light for interior spaces is apparent from the apron side of the terminal. J. Minimizing interior dividing and partitioning saves material consumption and maintenance as well as contributing to the light, airy feeling of the terminal. K. Durable, long-lived finishes like the locally quarried Texas granite ensure low maintenance and avoid wasteful replacement or refinishing over time.

F

G H

J

K

03 / THE PARADOX OF AMERICAN URBANISM

THE PARADOX OF AMERICAN URBANISM

In her landmark book, *The Death and Life of Great American Cities*, published in 1961, Jane Jacobs draws battle lines between two radically divergent visions of American urbanism. She decries "modern orthodox city planning" which had emerged in response to the advocacy of diverse figures like Ebenezer Howard, Sir Patrick Geddes, Frank Lloyd Wright, Le Corbusier, and others. Their vision of a healthy modern city integrated large quantities of open space into the built environment and allowed natural landscape to be prominent, if not dominant. Leading American urbanists in the mid-20th century, like Lewis Mumford, Clarence Stein, Henry Wright, and Catherine Bauer, had sought innovative patterns of land use and building form which embodied a social and cultural idea of living in a symbiotic, integrated relationship with natural open spaces.

Jacobs condemns this direction for its "monotony, sterility, and vulgarity," observing that, "Decades of preaching, writing, and exhorting by experts have gone into convincing us and our legislators that mush like this is good for us, as long as it comes bedded with grass." She offers quite the opposite vision for emerging American urbanism. She advocates the traditional city of streets, sidewalks, small blocks, tight building fabric and mixed-use. She declares that this formula will work because it has worked and should not be tampered with. Jacobs' critical voice was joined by contemporaries like Kevin Lynch, Gordon Cullen, and Edmund Bacon, all of whom admired, investigated, and learned from the traditional city and wrote influential books in the 1960s. Although certainly less vitriolic than Jacobs, their perspectives are nevertheless clearly critical of the emerging modern American city. Strongly influenced by this new generation of urbanists, the latter decades of the 20th century saw a somewhat begrudging return to lessons learned from European urban fabric and its reincarnations in the densest of 19th and early 20th century American cities.

In fact, though both of these polar visions outlined by Jacobs offer genuine and appropriate inspiration for contemporary American city-building, neither has been the kind of panacea envisioned by its advocates. American cities seem to be inherently paradoxical and resistant to pure solutions.

The landscape-dominated vision of Frank Lloyd Wright, Clarence Stein, and Lewis Mumford resonated in the wide-open spaces of the American Midwest, South, and West. It offered individual freedom of expression, a wide potential range of diversity in building form, and a specialization of land use that fit well with a legal system based on individual property rights and with a market-driven economy. It provided room for high-speed automobile travel and a decentralized lifestyle rooted in the single-family house. It empowered the individual as the fundamental unit of society.

But this landscape-dominated urbanism failed to fulfill other expectations of American society. In its non-hierarchical looseness it did become "mush," to use Jacob's term. It lacked identity—a "sense of place." Without more containment of space, there was a dearth of identifiable public gathering spaces. A collection of object buildings did not promote a sense of community and shared turf. The genuine paradox of American culture—its democratic reverence for the individual alongside an emphasis on community and solidarity—foiled the success of such a pure, clear vision.

The more traditional urbanism of Jacobs, Lynch, and Bacon likewise offered a powerful consonance with American culture in its creation of a network of continuous public spaces where citizens could exert a collective identity and where commerce became a significant force. Its creation of axes, monuments, and landmarks resonated with an American sense of pride and identity. It gave particular opportunities for cultural icons and emblems. This vision also fulfilled an American need to grow deeper roots. It advocated integration of old and new, rather than starting with tabula rasa.

But, this pure, vociferous vision failed in striking regards as well. It seemed inherently incapable of integrating the automobile without losing its most cogent assets. It defied a real estate market and land development economy that is inherently piecemeal—not holistic. It suggested a conformity and stifling of individual expression that often seemed forced, fake. The varied, even contradictory, nature of American political, economic, and social culture, again, defied such a pure vision.

Over the last two decades we have worked on a number of projects in downtown Austin wherein we have tried to deal with the paradox of American urbanism in a very particular and responsive way. We have sought inspiration from both poles of urbanism proffered in the last half of the 20th century, but have tried to develop a new hybrid from them which would be more useful in addressing real, tangible cultural and environmental needs.

Austin as a Case Study
Austin has offered an almost ideal context for this experimentation. It is a quintessentially American city. Founded in 1839 and conceived to be the Texas state capital from its inception, the city has a rich social, cultural, and political history. Its population has traditionally been well-educated and civic-minded with a keen awareness of the value of the physical environment and its profound effect on a community.

From its inception, the plan of Austin was conceived as a juxtaposition of powerful natural forces and a well-defined urban form. The site selected for the new capitol of the young Republic of Texas was chosen for its strong landscape character. Rooted at its southern end by the Colorado River, the site was framed on either side by two prominent bluffs about a mile and a half apart. At the base of each bluff was a

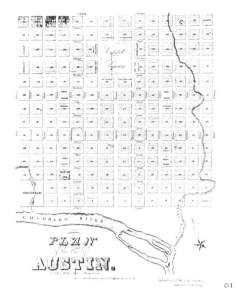

01 Waller Plan for Austin, 1839
02 Town Lake Park, Austin

creek—Shoal Creek at the foot of the western bluff, Waller Creek at the bottom of the eastern one. Between the two bluffs the ground rose gradually from the river northward to a hill located about equidistant between the strong edge features. The committee that selected the site immediately envisioned the capitol of the Republic on the hilltop, framed by nature's bookends on either side and facing down toward the Colorado River.

In 1839 Edwin Waller laid out the first plan of the city. He divided the east/west space between the two bluffs into 14 blocks delimited by 15 streets, the center one of which was named Congress Avenue. Wider than the rest, it aligned with the center of the hill intended for the capitol. The easternmost street was named East Avenue and the westernmost one West Avenue. The rest of the north/south streets were named for Texas Rivers—the westernmost being Rio Grande, the western boundary of the republic, and the easternmost being Red River, at the opposite end of its territory. In the north/south direction, Waller laid out the same 14-block dimension delimited by 15 streets. That meant the new grid stretched north of Capitol Hill a few blocks. The southernmost of these east/west streets along the river was named Water Street. The one on axis with the hill was made a bit wider, like Congress Avenue, and was called College Avenue. The rest of the east/west streets were named for Texas trees—Mesquite, Pecan, Live Oak, Peach, etc. (Later, the streets would become numbered.)

Four individual blocks in the grid were designated as greens or public places. These were evenly dispersed in the area between College Avenue and the river and on either side of Congress Avenue. The four blocks that occupied the top of Capitol Hill were bundled together in a much larger superblock which would provide a green setting for the Capitol Building. The 1839 plan of Austin neatly described a confluence between forces of nature—the river, the bluffs, the creeks, the hill, etc.—and a geometrical urban fabric—the 14-block by 14-block grid, the axes of Congress Avenue and College Avenue, the axial terminus of the Capitol Building, the geometrically spaced greens, etc. Even the grid of streets itself, the strongest of urban gestures, acknowledged its counterpart by naming its elements for natural features—rivers and trees.

The dialogue between natural form and urban geometrical form began almost immediately in downtown Austin and continues today. The grid never really managed to conquer the creeks. Bridges were not built because of expense or because of the threat of being washed out, so streets around the creeks became erratic and discontinuous. Eventually, along Shoal Creek to the west, the tight geometrical urban fabric implied by the grid became a loose series of green spaces—Duncan Park, House Park, and Pease Park. A similar pattern occurred to the east with Waller Creek generating Palm Park and Waterloo Park. The Jane Jacobs ideal of streets, sidewalks, small blocks, and tight building fabric got inevitably and beneficially compromised by the intervention of exactly what she decries—large quantities of open space allowing natural landscape to be prominent if not dominant.

The story of the development of downtown Austin is littered with examples of this paradoxical coexistence of two well-defined and warring camps of urbanism. It seems the populace, as well as the forces of both nature and economy, genuinely demanded a constant standoff between the two. In the planning and urban design work as well as in the architectural projects we have done in Austin, a recognition of the productive tension of this dichotomy—inherent in the earliest plans of the city and broadly evident in citizens' sentiments today—has been both inspiring and challenging.

The Town Lake Comprehensive Plan

The Town Lake Comprehensive Plan (TLCP) was done as a joint venture between Lawrence W. Speck Associates and Johnson, Johnson and Roy Landscape Architects and was completed in 1984. This two-year effort was the largest planning project ever done by the City of Austin and involved proposals for a 7.5-mile-long segment of the Colorado River running through the center of the city. The project included urban design recommendations for city fabric on both sides of the river as well as proposals for land acquisition and parks and public space development in the river corridor.

For over a century, the city had turned its back on the flood-prone river, lining its swampy banks with clay quarries, rail yards, a water treatment facility, two power plants, and a motley assortment of low-density warehouse buildings. Creation of Longhorn Dam in 1952 stabilized the river's water level and established Town Lake as a safe

Town Lake Comprehensive Plan for Austin, 1984

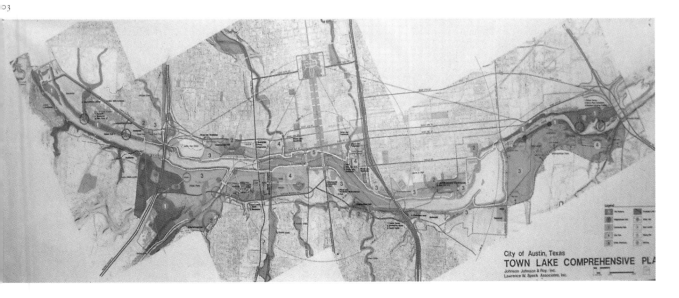

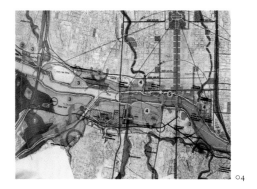

04 Town Lake Comprehensive Plan—detail
05 Town Lake Park Plan—detail
06 Town Lake Park Plan, 1984

and scenic body of water. While several extraordinarily attractive amenities developed prior to the dam were located in close proximity to the lake (Zilker Park, Barton Springs Pool, Deep Eddy Pool) the city had still not, by the early 1980s, managed to reorient itself to capture and consolidate a cohesive riverfront.

The new plan envisioned Town Lake as the heart of the city—a real urban hub where civic life and activity were concentrated. It imagined an urbanism fully committed to interaction and social gathering which might not be limited in physical form to the street, the square, and traditional urban parks. It conceived outdoor activities like walking, biking, boating, swimming, and sports as potential contributors to the urban scene. It integrated places for picnics, outdoor concerts, and casual dining with more high-brow cultural venues like museums, theaters, and performance halls. It offered opportunities for an urbanism which was both densely built and primitively natural in radical juxtaposition. It embodied a way of life that resisted pure stereotyping—that accommodated a broad range of citizen interests and perspectives.

The urbanism depicted in the TLCP embraces principles of both the modernist landscape-dominated attitude toward city-building damned by Jane Jacobs and her followers in the last decades of the 20th century, as well as tenets of a more building-dominated approach to creating city fabric which Jacobs and others advocated. It is comfortable with a weaving together of towers and other solitary buildings set in open space alongside tightly delimited urban rooms framed by well-coordinated façades of multiple buildings working together. It accepts the appropriateness of "one-off" object buildings as well as advocating, at times, the creation of highly regulated ensembles. It celebrates both individuality and the collective good. It balances the inevitability of a convenience-minded, car-oriented culture with the civility and gracious-

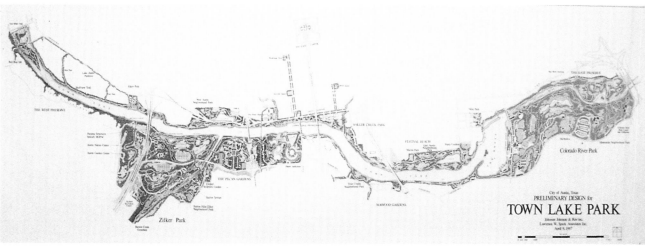

07 Town Lake Park—Hike and Bike Trail
08 Town Lake Park—Hike and Bike Trail
09 Barton Springs Pool

ness of well-designed pedestrian environments. It accommodates an economic and financial structure which is frequently focused on singular uses and markets, but, in the end, revels in mixed use by disallowing too much of any singular use in any one place and encouraging an overall balance of residential, office, commercial, and cultural/institutional uses.

One of the earliest steps in conceiving the TLCP involved outlining five different kinds of outdoor spaces we felt essential to include in the center of the city. Each was intended to add a dimension of experience of public space which would enhance overall life and activity.

<u>Preserves</u> were intended to locate, in the very heart of the city, some land that would be left in as natural a condition as possible. There was a desire not to exclude native flora and fauna of the region even in this area of densest human population. By retaining natural environments virtually inaccessible to humans, ecologies could be maintained which are essential in the larger balance of species interaction. It would not be likely, for example, that the largest urban bat colony in the world, which inhabits downtown Austin, would survive if there were not some substantial areas of dense vegetation and a full range of animal and insect species nearby. Carefully delineated preserves were therefore distributed throughout the length of the TLCP.

Sometimes the preserves were linked to research or education functions as in the case of land around the Nature Center or the large University of Texas Biological Research site (prime waterfront land which is committed to studying animal migratory patterns around the river). Other times, the preserve is made virtually inaccessible to anyone as in the case of some parts of Colorado River Park. Buildings and even trail improvements are pretty much excluded on preserve land.

<u>Neighborhood Parks</u> are intended to link Austin's thriving in-town neighborhoods directly into the life of downtown and Town Lake Park. One of the very healthiest features in central Austin is the presence of a series of well-established neighborhoods that ring the urban core. They are widely varied ethnically and demographically, and generally have houses from a range of eras—usually beginning in the 1920s or 1930s, but some going back to the 19th century. Of particular interest in the TLCP were Tarrytown, Clarksville, Rainey Street, Old West Austin, East Austin, and Govalle north of the river, and Bouldin Creek, Travis Heights, and Riverside south of the river.

The idea of the neighborhood parks was to create little enclaves that belonged quite solidly to the neighborhoods with their activities focused on the particular ethnicity or demography of local neighbors. Fiesta Gardens, for example, focused on celebrations and occasions particular to Hispanic and Mexicano East Austin adjacent to it. Stacey Park in Travis Heights was to link local pool and recreation facilities out to the broader Town Lake Park. Tiny Eilers Park in West Austin

10 Town Lake Park—Hike and Bike Trail
11 Town Lake Park—music event
12 Town Lake Park—Squeaky the Clown

was to provide picnic and playscape area for families living nearby but also was intended to create an outlet to the larger Hike and Bike Trail around the lake.

<u>Community Parks</u> were conceived as larger citywide facilities which would draw people from all over Austin to sophisticated recreation areas, sports venues, or individual activities. The Hike and Bike Trails were to link all the Community Park spaces together. Several spectacular icons were featured in the Community Parks. Barton Springs Pool, built in the 1930s in Zilker Park, sets a benchmark for great urban recreation spaces with its distinguished pool house and scenic sunning slopes. Deep Eddy Pool creates a mecca for family recreation with its enormous children's pool as well as its serious sports swimming area nestled under towering cottonwood trees.

Part of the idea of these Community Parks was that in Austin the kind of "seeing and being seen" that occurs in Italy on the piazza or in Mexico on the plaza might happen here on the Hike and Bike Trail. Teenagers "scoping each other out" might occur just as likely in the promenade around Barton Springs Pool as in a kind of café society on a boulevard one might have found in a traditional European city. The same kinds of community-building urban places that create a powerful downtown identity and sense of place might be equally potent in a fresh and more particular kind of "urban" setting here.

The attitude in all Community Parks was to take a lesson from Barton Springs and Deep Eddy Pools and to emphasize scenic aspects over the creation of just "facilities." Lighted sports fields and spectator stands were relegated to a very few areas, keeping the predominant feeling soft, green, and park-like.

<u>Cultural Parks</u> were intended to be settings for activities that would, in a strictly Jane Jacobs kind of urbanism, be located on the square or terminating the axis or spilling onto the city streets. But in Austin it seemed like cultural institutions such as a performing arts center, a community events center, a Mexican-American cultural center, a community arts center, a nature center, a science museum, etc. might just as appropriately occur in a symbiotic relationship with an active green space as with a street or plaza. These kinds of institutions were seen as opportunities to bleed the boundary between recreation and cultural activities in the form of outdoor concerts, outdoor musicals, music festivals, and even fundraisers for community causes like the Capitol 10K run.

Though there was a strong commitment to some cultural events like the Paramount Theatre being on a main street like Congress Avenue or the Austin Museum of Art being on a public plaza like Republic Square, there was also a sense that having cultural institutions in a wide range of settings could enrich the cultural life of a city. A wider range of events might be germinated and a broader range of people

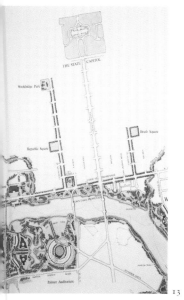

might get involved. Buildings to house cultural events in a park context were imagined to be pavilions in open space, giving a range of opportunities for architectural expression. (It is sad to see institutions trying to make their mark by plopping "object" buildings in "fabric" situations. Isn't it better to have some places where "object" buildings make sense?)

The life of a cultural park would be urban, but informal. There would be mixed use, but the uses might be dog-walking, picnicking, soccer games, and antique shows, rather than the more conventional retail, office, and residential uses. This pattern acknowledges the broader range of late 20th-century/early 21st-century life and the powerful role of leisure and recreational activities in our lives—especially in a city like Austin.

<u>Urban Waterfront</u> is the designation given to open spaces that were to be strongly defined by building edges. This is where the traditional realm of street, sidewalk, and square resides—where spaces are closely contained. In the district of the TLCP this was primarily in the stretch where Town Lake Park runs through the original city grid. Here, the notion was that a strong architectural edge should be created which would be of sufficient scale to create a kind of "room" of park space but not so gigantic as to dominate the river and park edge. This boundary would best be relatively continuous, taking a cue, perhaps, from the extant scale of the Seaholm Power Plant (a beloved and powerful relic just west of the point where Shoal Creek empties into Town Lake). There should not, it was recommended, be any more towers at this edge.

This strong urban boundary was intended to create a striking dialogue between the tight, gridded urban fabric of downtown and the lush, green, natural park space. The green of the park space was meant to filter into downtown via five routes. Shoal Creek and Waller Creek would draw loose meandering park space into the grid eroding it and providing interesting sites for opportunities not available elsewhere in the tight street fabric. In addition, two of the north/south streets, Guadalupe Street on the west side and Trinity Street on the east side, were designated "green fingers." Though they would maintain their linear, geometric form, they would get more street trees, more pocket parks than neighboring streets. They were selected both for the fact that they ran uninterrupted from the waterfront deeper into the city than their neighbors and for the fact that they ran alongside the four blocks originally designated for parks in the 1839 plan. Finally, the fifth penetration of the park presence into the tight downtown fabric would be Congress Avenue—a very broad, tree-lined processional culminating in the lush parkland around the Capitol Building.

All of these open space types—the Preserves, Neighborhood Parks, Community Parks, Cultural Parks, and Urban Waterfront—contribute a particular dimension to public life in the city. They also exercise a full breadth of relationships between buildings and the

3   Town Lake Park Plan—downtown waterfront development
4   Town Lake Park—waterfront

15 Congress Avenue, Austin, Texas
16 Sixth Street, Austin, Texas

natural environment. Much of the effort of creating the Town Lake Comprehensive Plan went into locating where what open space type might be appropriate through the 7.5-mile length of the Town Lake Corridor. As with all such planning efforts, there is no such thing as full implementation, but the TLCP has made a striking mark on the city in the 20 years since its adoption, not least in several landmark building projects that we have been involved in which follow.

Austin Convention Center

Three years after the completion of the Town Lake Comprehensive Plan we were hired by the City of Austin to do final site designation and design for a new Austin Convention Center. Though a general location for the convention center in the lower part of downtown had been suggested in the TLCP a great deal of work was required to outline a quantity of land and exact location which would not only serve the initial needs of the convention center, but could also accommodate future expansion.

The commitment to place the convention center in a neglected part of downtown was critical. The project was intended to seed downtown development. The site eventually chosen reinforced the commitment on the part of the city's hotels to stay downtown by locating near them. It significantly boosted a flagging Sixth Street Entertainment District, stimulating it to become one of the most valued assets of the community. And, in just over ten years, it helped transform over 20 blocks of derelict buildings and open parking lots into a district with not only the original and expanded convention center, but also two large new hotels, several renovated and new multi-family housing projects, and a dozen or so restaurants—mostly in renovated buildings.

The southeast corner of downtown where the site was located was an eclectic mélange of disparate elements. Two blocks east of the site was Interstate Highway 35, a mammoth elevated freeway which was cut through downtown Austin in the 1960s. Two blocks southeast was the Rainey Street neighborhood, a small but sweet and neatly preserved enclave of tiny single-family houses on tree-lined streets. Two blocks south of the site was one of the very nicest parts of Town Lake Park where Waller Creek joins the river. The bluffs were dramatic, the banks wide, and the trees huge in this part of the park. Two blocks to the west were high-rise office towers on Congress Avenue. Two blocks to the north was the Sixth Street Historic District, struggling to become an entertainment focus for the city.

The immediate surrounds of the site were also strikingly divergent. The four-block tract was bounded on the south by Cesar Chavez Street, a major east/west connector and something of a ceremonial street in the city. On its west was Trinity Street, a thoroughfare with little automobile traffic but which had been slated in the Town Lake Comprehensive Plan to become the Trinity Street Green Finger. It was

7 Convention Center District Plan, Austin, Texas
8 Austin Convention Center

intended to make a strong pedestrian connection between the Sixth Street Entertainment District to the north and Town Lake Park to the south. The northern boundary was Third Street—another street with very little traffic which, in fact, was planned to be closed when expansion occurred. The eastern edge of the site was Red River Street which had excellent north/south connectivity for vehicles but little actual traffic. Across Red River was an enclave of tiny historic wooden buildings and Palm Park. Clipping the southeast corner of the site was Waller Creek with its steep limestone banks and big mature trees. Across the creek was a historic building which once housed Weigl Iron Works, a source of extraordinary German metal craft in the early 20th century. The little metal building had for many years been a popular barbeque restaurant.

The building type we were placing on this complex and provocative site is not one generally considered to be urban-friendly. As was noted in an article in *Progressive Architecture* shortly after the Austin Convention Center was complete, these institutions are generally characterized by "vast exhibition/meeting rooms, smaller multi-purpose rooms, and prefunction areas, jammed into featureless, windowless volumes devoid of relation to their surroundings."[13] The *PA* article went on to describe our very different approach:

> Here, in a new building on a site open to all sides, yet oriented to different urban and landscape conditions, one can articulate the parts and put them into dialogue with their surroundings.... We see that a convention center need not be one great mass, but rather something of a village in itself. Spaces are differentiated both with reference to their own use or place in the internal organization as well as in relation to site conditions. These changes in form are paralleled by changes in materials. Yet this appears not as a nostalgic or historicizing "village," but as a genuine rethinking of the organization of large systems in elements that consider the person and the more intimate scale of the urban landscape.[14]

We placed the largest-scaled spaces, the exhibition halls, in the center of the site and wrapped them with the "village" of smaller-scaled functions facing the streets and the creek around the perimeter. Each programmatic element in the "village" finds its own shape, form, and material treatment in response to its function and location. Various parts of the building honestly express the broad range of activities that occur within. Lobbies, stairs, escalators, etc. are allowed to take on their own individual identity on the façade as well. As all of the various elements add up, they create a streetscape which is similar in richness, scale, and diversity to what one finds on nearby Congress Avenue or Sixth Street. The kind of articulation that might have been created by a

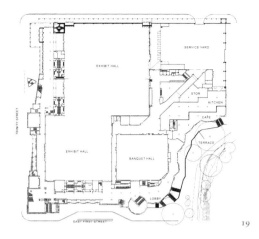

19 Austin Convention Center—first floor plan
20 Austin Convention Center—vest-pocket park on Trinity Street

drugstore next to a haberdashery next to a hardware store is not so different in general character from that created by a lobby next to a meeting room next to a prefunction space. Without being cloying or condescending, the building becomes a natural extension of the existing city.

Various functions of the convention center are matched with varied conditions on the site. The more relaxed functions of the café, bar, banquet hall, lobby, and terraces are stretched along the more casual edge of the site—along Waller Creek. This part of the building is loose and faceted, creating a relaxed feeling appropriate to the functions. It skips in and out, reacting to the mature trees and the bend in the creek. The building is low in scale here, stepping gradually up from creek bed to terrace to lobby to banquet hall. It does not overpower the scale of the landscape or the little historic buildings around. Materials are mostly a random ashlar stone, which is sympathetic to the shapes, colors, and textures of the limestone creek bed.

As the building transitions from the Waller Creek side to the Cesar Chavez Street façade, the loose facets of the terraces become more regular, forming a tall polygonal lobby. This strong feature anchors the east end of the Cesar Chavez Street façade and is roughly matched by a similar rectangular lobby on the west end. Here, the most ceremonial functions of the building find their home on the most visible and ceremonial street in the district. The polygonal lobby becomes the primary "front door" to the banquet hall while the rectangular lobby serves a similar function for the exhibit halls. Between the two landmarks, a bank of escalators is marked by a notably vertical expression in the façade, and two-story prefunction spaces are lined by generous porches where occupants can spill outdoors for breaks between events. The porches also double as effective sun shades on this southern face.

The Trinity Street façade expresses the more everyday "workhorse" functions of the building. Various meeting rooms each receive individual treatment. The most elaborate of these is placed at the terminus of Second Street so that it gets an unobstructed view west to the Hill Country. Its large oculus-shaped window with a carefully articulated sun shade becomes a landmark in the city when viewed down Second Street. Next-tier meeting rooms occupy simple masonry volumes and have large rectangular openings onto the pedestrian-friendly Trinity Street. The more generic and subdividable meeting spaces are in a raised metal-clad box that reflects their rhythms and repetitiveness honestly. Escalators and stairs on the Trinity Street façade are marked by appropriate stepped and vertical volumes. Prefunction spaces have similar double-decker porches to those on the south façade.

The building line is varied along Trinity Street creating a series of tree-filled vest-pocket parks as a means to reinforce the intended "green finger" character of the street. Because transit stops are located here and because the convention center parking garage is a block to the west, this is a common place for everyday arrival to the building.

The Terrace Offices—site
The Terrace Offices—entry court
The Terrace Offices—east façade

The gentle, almost completely permeable nature of the building here contrasts with the more ceremonial approaches off Cesar Chavez Street. A deep covered loggia at the north end provides a sheltered place for meeting, greeting, or waiting for transportation.

The Austin Convention Center is marked by the genuine paradoxes of American urbanism. It is a traditionally "urban" building with façades creating edges to well-defined street spaces. It makes good city fabric. But it is also a building set in nature in some ways, reveling in the loose, enriching intrusion of Waller Creek on the downtown grid and drawing some of the lush greenery of Town Lake up Trinity Street into the city. It takes the freedom to acknowledge the disparate, sometimes chaotic nature of the American city which is often more about precipitous shifts and mind-boggling contrasts than about clarity or consistency. As the previously mentioned *PA* article notes, this is a "building that can set a pattern for civic development, maintaining the scale of Austin's historic fabric while projecting messages not only about the city's divisions but about its ultimately invaluable diversity."[15]

The Terrace Offices

Located about two miles south of the Town Lake Corridor, the Terrace Offices are perched on the side of a bluff that looks back to downtown. This project deals with an increasingly prevalent paradox of American urbanism—a medium-density office complex nested in a dominantly natural setting. In this case, the site was an ecologically sensitive, never-developed tract adjacent to a nature preserve. Because of its proximity to downtown, however, development pressure was significant.

The American love affair with the outdoors and our desire to live and work in close proximity to nature can sometimes create ironic situations. The permitted density on this beautiful, unspoiled site could easily have obliterated the very assets that made it especially appealing to begin with. The master plan originally approved by the City of Austin for site development would have created a fairly normal office park with generic boxy buildings straddling precipitous topographical changes with plenty of manicured lawns. It would have destroyed many of the mature trees on the site and exiled the kind of flora and fauna that had so amiably inhabited it.

The revisions we suggested included stringing the buildings' long and thin parallel to contour lines and minimizing cut-and-fill. By bending the buildings slightly, we could save all of the real specimen trees and retain much of the existing topography. We also made a very strong point of leaving the "thicket" of vegetation on the site intact insofar as possible. During construction, fences were erected ten feet outside the building perimeter to keep construction spoilage to a minimum. A price tag was put on all the adjacent trees so that if one was destroyed or damaged fines could be assessed. The goal was to place this relatively dense urban element on the site with as little disruption as possible.

24 Computer Sciences Corporation from Town Lake Park
25 Computer Sciences Corporation—south façade

Impervious cover on the site was kept to a bare minimum. Access roads were made the most direct and efficient possible. A single paved area was designed to triple function as access to the garage, drop-off, and turn-around for service vehicles. The little court was even walled off from adjacent native landscape to prevent intrusion. Openings in the wall give cues that this natural environment is to be enjoyed visually, but not disturbed.

The result is a rather startling urban gesture—a powerful juxtaposition of built and native environments. Suburban accoutrements like lawns, shrubs, ornamental landscaping, and planting beds are absent. The vocabulary is tight, clean, and urban on one hand and wild, native, and natural on the other. The architectural vocabulary reinforces this notion. Stone walls begin thick and dense with small openings looking out into the near-view thicket on lower floors. As the building rises and the thicket becomes less dense, the windows increase in size inviting longer views. The top floors are almost all window, the stone piers having diminished and converted to precast concrete, inviting dramatic views across the valley to downtown.

Computer Sciences Corporation

When Computer Sciences Corporation (CSC) was making plans to create a large new facility for its operations in Austin, their first inclination was to look for a suburban site at the edge of the city. CSC, like most other software companies, had a history in its operations worldwide of locating in "office-park" or "campus" environments. Their expectation was to build a secure (even gated) precinct wherein several buildings and a parking structure would be set in a single-use, landscape-dominated context.

This sort of thinking about site selection had become commonplace among other high-tech companies in Austin, such as Dell, Motorola, and IBM. As these knowledge-based corporations expanded, they tended to produce multi-building complexes on the fringes of the city. The subsequent roads, utilities, housing, and other services that were required for their high-tech workforce were sprawling deep into the Texas Hill Country causing justifiable alarm among environmentalists and taxing both governmental and private sector resources as they scrambled to provide infrastructure for rapid growth. This tendency provoked some serious soul-searching as to Austin's image of itself. The impact of rampant urban sprawl with its associated traffic congestion, air and water quality degradation, and destruction of valued scenic landscape posed a fundamental lifestyle threat in Austin and provoked vociferous political debate.

In the meantime, Austin's downtown was enhancing its role as home to a vibrant arts and entertainment scene, but had stagnated somewhat as a home for businesses. The presence of the convention center and most of the major hotels downtown had populated the

26 City Hall District-model, Austin, Texas
27 Computer Sciences Corporation—detail

urban core with visitors anxious to bolster Austin's vital club and live music scene. In-town neighborhoods provided enough nearby residential support to maintain dozens of good restaurants and even an upscale health food grocery. Most of the local arts organizations were committed to locations downtown drawing their patrons from all over the city. Town Lake Park had become a kind of "living room" for the metropolitan area attracting thousands of daily users to the Hike and Bike Trail and tens of thousands to special events and festivities. But there were parts of downtown where nothing happened above ground level—where density of people and activity was sorely lacking.

One of the most visible areas of this sort was a five-block district in the southwest part of downtown where the City of Austin owned most of the real estate. Mayor Kirk Watson, elected in 1997, was fond of saying, "If you asked an Austinite to walk through the downtown and point out its ugliest four or five blocks, these would be the ones." He was committed to constructing a new city hall in this district, consistent with the Town Lake Comprehensive Plan, and to encouraging development of a mixed-use district around it.

For many months there were ongoing discussions and negotiations between the City of Austin and CSC to see if this district might be a suitable spot for the 600,000 square feet of new office space CSC was planning. As architects for CSC, we participated fully in projecting and testing a range of possibilities for matching the very real needs of an industry based on suburban notions of office development and a downtown core hungry for an infusion of life and people who were a part of the "new" Austin economy. There was talk of a "digital district," where CSC would be a key stimulus with housing, retail, governmental, and arts facilities working together to produce a physically and culturally rich urban ensemble.

The scheme that was eventually selected in a joint agreement between the City of Austin, CSC, and a large residential developer created two blocks of offices with retail on the ground floor for CSC, two blocks of multi-family housing with retail on the ground floor for the developers, and one block for City Hall and its plaza. City Hall would occupy the focal point at the end of Drake Bridge with its plaza toward Town Lake Park. CSC would occupy the two blocks on either side of City Hall, creating a frame for the public plaza and what was hoped would be an iconic civic building. The multifamily housing would occupy the two blocks behind the CSC buildings on Second Street. Retail on the ground floor of all of the buildings would orient strongly toward Second Street, which would become a new pedestrian-focused retail core at the heart of the district.

The City of Austin committed substantial funds to infrastructure improvements (though still far less than it would have cost to produce suburban infrastructure for this size development). Their incentives included a "Great Streets" program to refurbish all streets and side-

28 Computer Sciences Corporation—detail
29 Computer Sciences Corporation—detail

walks in the district and the provision of a central heating and cooling plant which would provide substantial energy savings. Even with such incentives, however, this was a difficult and rather daring commitment to make on the part of CSC. They had to be convinced that they could get all of the amenities they were accustomed to in a suburban location but with some value added for being in the heart of the city.

Three perceived assets of suburban locations that had to be captured on this urban site were security, ease of access (particularly by automobile), and views. Because there is a great deal of proprietary information involved in the work CSC does (both in terms of CSC's own research and development tools and in terms of their client's data), it was critical that a secure work environment be maintained. This was important not only within a single unit or building, but within the entire complex. Data, equipment, and machines had to be able to move freely around the whole facility without a breech of security. On a suburban site this was generally accomplished by creating a secure compound that could be gated for privileged entry.

On the downtown site we solved this problem in two ways. First, we made the lobbies far more important places than they would normally have been. They became, not only places for screening people for entry, but also spaces where interactions with people from outside could occur in a non-secure environment, thus avoiding screening. It was critically important, however, that these lobbies not seem like heavy-handed security barriers. Just as the drive along a curving wooded road makes a guard station seem less threatening to the authorized visitor in a suburban environment, so something had to be done in much less space to accomplish the same civil and gentle feeling of security here.

A second measure taken to achieve an internal secure environment was to build a tunnel connecting the two buildings a block apart from each other. Intentionally not convenient for everyday movement of workers from one building to the other, the tunnel provides only for transfer of secure documents, equipment and machines. All other movement between the two buildings actually benefits from and contributes to the urban street-scene of the district.

Ease of access to and from the site, especially by automobile, had to be accomplished at least as easily as on a suburban site. There was to be a great deal of coming and going to the buildings at all hours of the day by CSC employees, who travel a great deal and have frequent meetings with clients and others off-site. Parking areas needed to feel safe and routes from parking to work spaces needed to be clear, simple, and direct. The solution we chose brought the parking directly into each building via a secure entry portal. Access could then be gained directly into work space via card-key without going up or down in elevators. This approach actually addressed CSC's concern much better than is done on most suburban sites. The density and compactness as well as the higher price of urban land provoked a more efficient solution with smaller travel distances.

30

31

30 City Hall District from Town Lake Park
31 Schneider Store and Computer Sciences Corporation

Many suburban sites in Austin are blessed with excellent views in virtually every direction. For many high-tech workers, a view to a serene natural environment from one's workspace is a prized asset. It was important to prove to CSC that we could match the view potential of a suburban site in a downtown environment. Given the availability of Town Lake Park on one side of each site we were able to maximize views in this direction and thus gain the sense of distance, big sky, greenery, and water which is so treasured in Central Texas.

In addition to capturing these assets commonly available at the edge of the city, CSC understood that they got substantial bonuses for their employees by being downtown. They could participate in a lively mixed-use environment which offered options for lunchtime or after-work activities far richer than what is available in the suburbs. There would also be distinct options for living and working downtown and avoiding commuting altogether—even within the "digital district" itself. But it is important to understand that the kind of "Jane Jacobs urbanism" described earlier was not compelling enough to bring CSC downtown on its own. It was a distinctly new kind of urbanism strongly incorporating parks, recreation, and open space that attracted CSC back to the core of the city. It was the accommodation and taming of the car that made this kind of urban environment work for them. It was the acknowledgement of new business and security realities very different from those that shaped traditional cities that made a downtown location make sense for this emerging industry.

It is fundamentally important that these issues be engaged and that we find a way to attract the industries that have fled to the suburbs back to the city center. It is romantic and nostalgic to think that the ingredients that went into a 19th- or early 20th-century core are going to recreate themselves in the 21st century. As industries, businesses, and economies change, so must urbanism find new forms to reflect new cultures. One of the most exciting things about some younger (and growing) American cities like Austin is the opportunity they represent to create new urban forms and new urban social chemistry.

The architecture of the CSC project emerges overtly from its urbanism. The buildings have solid, simple massing reflecting the powerful grid of downtown. Their six-story height is intended to create a clear edge to Town Lake without overwhelming the scale of the landscape—just what was stipulated in the TLCP. They create a frame for the axis of Drake Bridge and bookends for the new City Hall and Plaza.

The westernmost building is notched on one corner to preserve the only significant preexisting structure on the site—the mid 19th-century J. P. Schneider Dry Goods Store. The simple rectangular volume of this historic structure, typical of what was once found in much of downtown Austin, is made of a local buff-colored brick laid up with elegant and straightforward tectonic clarity. The CSC buildings take some of that same tectonic clarity and apply it at a larger scale to

32 Computer Sciences Corporation—lobby

contemporary construction technology. Hefty masonry piers made of load-bearing Leuders Roughback stone create a general exterior armature for the buildings. The Roughback stone is used in large blocks and revels in the wide range of color variation—from creamy buff to caramel—inherent in its geological formation. Horizontal precast spandrels tie the piers together on the lower floors.

Placed substantially back from the face of the masonry, an intricate glass curtainwall interlocks with the stone piers. The glass is set in two different planes to give relief and to emphasize its thinness. A complex layering of colors in the glass—from milky white to a deep watery green—animate the façade further as well as help to ameliorate visual differences between lower floor-to-floor heights in the garage and taller ones in the office areas. A copper sunshade of rather heroic scale helps to terminate the piers and readdress the horizontal line of the river's edge.

These buildings are robust urban fabric. As commercial structures, they make no claim to the kind of object quality that, in this district, needed to be reserved for City Hall. They are constructed of materials that integrate them inextricably into their surroundings. The Leuders limestone, by its color and texture, creates a rich, warm dialogue with both the limestone bluffs of the river nearby and the historic remnants of old downtown Austin. The deep green glass has the same reflectivity and color as the river's surface. But both stone and glass are also very particular and identifiable as themselves. They are not neutral or anonymous. They have their own distinct character.

Even the interiors at CSC owe much to the urban context. A small courtyard at each building's entry, facing Town Lake, pulls a bit of the park space onto the site. The entry lobbies extend that gesture with their backlit glass walls that draw the shimmery lightness of the river right into the building. These large, carefully programmed spaces reinterpret the loose, interlocking spaces of the park in sumptuous, tactile materials—several kinds of greenish glass, *stucco lustro* piers in deep, rich colors, intricately grained wood. The lobbies are inspired by the park space, but do not mimic it literally. They just have a feel that is compatible.

Upstairs spaces are all about the view with 13-foot windows at the perimeter that draw the eye outward. Open plans and glass interior partitions where rooms are required, bring the sense of outdoors deep into the building. Special facilities like lockers and showers facilitate employees' taking advantage of the Hike and Bike Trail and other recreational amenities of Town Lake Park.

As *Architectural Record* noted in a feature shortly after the buildings were completed, "The strength of the CSC complex is its urbanity. Its buildings were conceived as parts of a civic and commercial landscape, not as discrete, grandstanding objects. They do not strut or preen.... In an age of steroidal architecture, you have to applaud that." The CSC buildings advocate an urbanism that reflects the paradoxes of the

Austin City Lofts from Fifth Street

contemporary American urban scene naturally and sincerely. They are a genuine attempt to make architecture grow out of larger concerns for the city and thereby make it a tool for the creation of an appropriate and meaningful urbanism.

Austin City Lofts

As early as May 1984, in a planning study done by Denise Scott Brown, the lower part of Shoal Creek, near its termination into Town Lake, was proposed as a high-density residential district. The Town Lake Comprehensive Plan reinforced that direction. Ideal for multi-family housing, the district is bounded to the west by an area containing shops, restaurants, health clubs, and a large grocery store, rare for downtown environs. Just to the east is Austin's central business district with its requisite office towers, but inhabited at much of the street level by a lively entertainment district. North along Shoal Creek is the two-mile-long, meandering Pease Park that stretches up into several of the most sought-after in-town neighborhoods in the city. Just to the south is the seven-mile-long Town Lake Park with excellent recreational amenities.

This is an almost ideal context in which to weave together the landscape-dominated urbanism with towers placed in large open spaces, and the more building-dominated city fabric with well-defined streets and contained spaces described in the early part of this essay. The site for Austin City Loft's, a 14-story, 82-unit multi-family residential building, is just at the edge of the original Waller grid and located on Fifth Street, a major arterial entering downtown from the west.

We chose to tie the building strongly to the downtown grid by aligning it with the Waller Plan, even though Fifth Street skews a bit as it extends off the original city plat. We emphasized the containment of the street with a three-story, load-bearing masonry wall, its height as well as its materiality matching the pedestrian scale of much of downtown. Retail space on the ground floor as well as the project's entry lobby lines the Fifth Street façade.

The building demurs to the landscape character of Shoal Creek immediately to the west. It sets back from the waterway and Hike and Bike Trail to create a pool and recreation area with an outdoor entertainment space. The scale is broken down by varying the volumes of the building and by utilizing a range of materials and treatments—stone, copper shingles, concrete, corrugated metal. This edge of the site, which is a full story below street level, is soft and green. Huge tree canopies dominate vistas. The building becomes a tower set in a park.

Even inside, the design of living spaces revels in this duality of landscape and city, openness and containment. Lower units are nestled into treetops. Upper units have spectacular views both to the city and to the hills and the greenbelt of Town Lake. Everyone has generous outdoor spaces—terraces, outdoor rooms, and balconies. Life here embraces the vital, enriching paradox of American urbanism.

88 / AUSTIN CONVENTION CENTER

A

B

A. Each programmatic element on the Trinity Street façade finds its own shape, form, and material treatment in response to its function and location. **B.** As all of the various building elements add up, they create a streetscape which is similar in richness, scale, and diversity to what one finds in the best areas of downtown Austin.

Overleaf: **C.** A tall, polygonal lobby anchors the east end of the Cesar Chavez Street side of the building, creating one of several orienting landmarks along the center's perimeter concourse. **D.** The polygonal lobby creates a transition on the building's exterior between the loosely faceted creek edge and the more ceremonial façade on Cesar Chavez Street.

C D

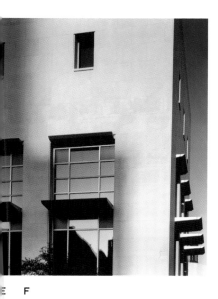

E. East façade on Waller Creek steps gradually up from creek bed to terrace to lobby to banquet hall employing irregular forms which react to locations of mature trees and the bend in the creek. F. The pavilion at the corner of Cesar Chavez and Trinity Streets is faced with smooth cut local limestone befitting its landmark urban position. G. A major entry point at the northwest corner of the site is marked by a deep, open loggia with a suite of meeting rooms above, faced in aluminum panels. H. Concourses between the building's landmark pavilions are faced with generous porches which provide effective sun shading as well as outdoor spaces where occupants can spill out for breaks and events.

E F

G H

94 / COMPUTER SCIENCES CORPORATION

A

C D E

Previous page: **A.** The architecture of the CSC project emerges overtly from its urban situation with solid, simple massing that reflects the powerful grid of downtown Austin.

B. The westernmost CSC building is notched on one corner to preserve the only significant historical building in the district-the J. P. Schneider Dry Goods Store. **C.** The new buildings take some of the tectonic clarity of the Schneider Store and apply it at a larger scale to contemporary construction technology. **D.** The local buff-colored brick of the Schneider Store is replaced by piers and walls in the CSC buildings made of load-bearing Leuders Roughback stone. **E.** Stone with precast concrete spandrels is joined with an intricate glass curtain-wall and a copper sunshade to create the building's façades.

 F
 G
 H

F. A small courtyard at each building's entry, facing Town Lake, pulls a bit of the park space onto the site and even into lobbyscapes. G. Lobby interiors respond to the reflections and shimmery surfaces of the lake as well as focusing views outward to the waterfront. H. Sumptuous, tactile materials line every face of the lobbies-several kinds of greenish glass, Venetian plaster piers in deep rich colors, intricately grained wood, polished limestone floors. I. Glass walls in the lobbies are back-lit, creating an indeterminate boundary to the rooms as well as providing general illumination. J. The lobbies are inspired by the lake and the park space, but do not mimic it literally. They just have a feel that is compatible.

100/
AUSTIN CITY LOFTS

A. The scale of the 14-story residential building is broken down by varying its volumes and utilizing a range of materials-including stone, copper shingles, concrete, and corrugated metal. **B.** North of the building a two-mile-long greenbelt stretches along Shoal Creek, providing long vistas into parkspace as well as direct access for recreational uses. **C.** Local city fabric is diverse and lively with an eclectic mix of restaurants, bars, clubs, shops, and even a fine large grocery store.

Containment of the street space on Fifth Street is emphasized with a three-story, load-bearing stone wall, its height and materiality compatible with the pedestrian scale of much of downtown. **E.** Even in the elevator lobbies, connections back to the surrounding city and greenbelts are emphasized with generous windows. **F.** The west façade of the building sets back from Shoal Creek creating a pool and recreation area with an outdoor entertainment space.

E

F

G. Unit interiors mix structural concrete and steel with finer level of finish.
H. Rooms are strongly oriented to views through large south-facing windows with deep horizontal sunshades. I. Multicolored panels incorporate kitchen cabinets into walls that define entry, bathroom, and storage closets.

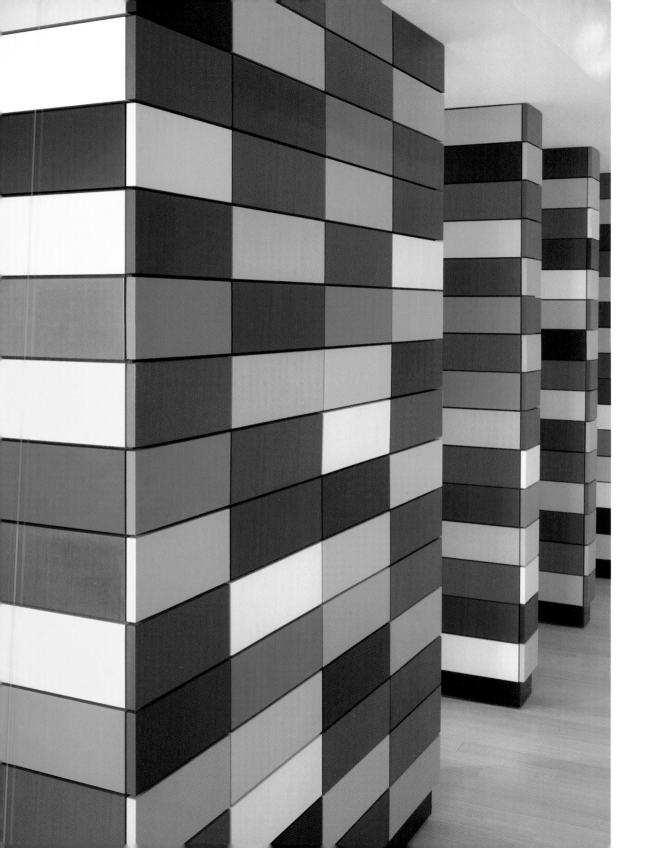

J. Double-height living room spaces face south. **K.** Deep terrace connects living room space to the city and Town Lake. **L.** Study overlooks living room with duplicate view to city and Town Lake.

04 /

TECHNOLOGY AS A SOURCE OF BEAUTY

TECHNOLOGY AS A SOURCE OF BEAUTY

In the spring of 2003 I was invited to co-teach an interdisciplinary course on "beauty" in the honors liberal arts program at University of Texas at Austin. Because I view architecture very much as an interdisciplinary field—part art, part engineering, part business, part philosophy, part sociology, part political policy, etc.—I treasure such opportunities to delve deeply into issues relevant to architecture from perspectives outside the field. This particular course involved seven faculty members—an expert in constitutional law, a mathematician, a poet, a cosmologist, a scholar of classical philosophy, an economist, and an architect.

Though there were many definitions of beauty and constructs of what is involved in creating beauty proffered in the course, one particular discussion struck me as by far the most fruitful in stipulating overarching criteria for evaluating beauty in a very wide range of fields. The discussion was led by the legal scholar of the group and focused on what created beauty in the law. He made the point that, above all, a beautiful legal argument relied on craft. It had to be carefully constructed, elegantly assembled, and honed with meticulous care. The interdisciplinary group embraced the notion of craft immediately as an applicable criteria for beauty in other fields. A finely crafted couplet in poetry is beautiful. A finely honed sonata in music is beautiful. A precise, refined mathematic proof is beautiful.

Our legal scholar went on to outline three components that seemed to best define craft in the context of a legal argument. First, a lean and direct economy is essential. Nothing can be superficial or extraneous. Second, there must be a rich complexity with subtlety and nuance informing every element. Third, there must be potency. The argument must make a difference, result in consequences, have an impact. Like the initial notion of craft, these three components—economy, complexity, and potency—were also endorsed by the interdisciplinary group as having real resonance and contributing fundamental beauty to our various fields. The cosmologist acknowledged that it is the economy, the simplicity, the uncontrived nature of the solar system that lends it such incontrovertible beauty. The poet defined beauty as "using all of the tools in the toolbox"—incorporating richness, texture, layering, complexity. The philosopher observed the potency of things beautiful. They make us feel, make us yearn, make us care.

Of course the architect in me delighted in these discussions. I was thrilled to hear a word like "craft," so much a part of our lexicon as builders, deemed to be so central to the notion of beauty. It was inspiring to make the connection between the hard work of collecting, assembling, editing, and detailing and the creation of things beautiful. It reminded me how much the generation of beauty in architecture has to do with the act of making, the art of construction, the deployment of materials, the implementation of technology.

1 Machu Piccu, Peru
2 Sachsahuaman, Cuzco, Peru

Elaine Scarry, in her recent book, *On Beauty and Being Just*, notes that, "though the vocabulary of beauty has been banished or driven underground in the humanities for the last two decades, it has been openly in play in those fields that aspire to have 'truth' as their object—math, physics, astrophysics, chemistry, biochemistry—where every day in laboratories and seminar rooms participants speak of problems… that are 'beautiful,' approaches that are 'elegant,' 'simple.'"[16] Here again, the search for what is real, what is tangible, what is clear, what is incontrovertible is seen as a source of beauty. Scarry speaks of an appealing "aspiration for enduring certitude." Is there a timeless sense of beauty that is tied to the "truths" inherent in science and technology? Is there a path to the creation of beauty that requires a constant search into what is real, honest, true in the physical world? How might this apply to architecture?

I will confess that a great deal of my own love of architecture, of my own thrill in its beauty, comes from an appreciation of elegant craft, construction, material deployment, and technology. These seem to me to be eternal verities of architecture—part of what elevates it to fine art. I am awestruck at the amazing beauty of places like Machu Piccu in Peru, where the architecture is not about exotic shapes or forms, not about fancy or pretentious materials, but rather derives an extraordinary presence simply from the art of building—of laying habitable spaces into the landscape, of quarrying stone, of stacking it up into walls and openings.

It is the craft of the architecture here that makes us awestruck. It is the technological prowess of the Inca builders, not only at Machu Piccu, but perhaps even more impressively at nearby Sachsahuaman, Oyantaitambo, and Pisac, that creates timeless beauty. The three elements mentioned earlier—economy, complexity, and potency—are certainly operative here. Formal gestures are subtle and economical, but they are also varied and complex. There is a clear, honest understanding of this particular local stone—its density, its capacity, its texture and surface. Architecture emerged here from a simple, elegant expression of technology centuries ago, but maintains a beauty and poetry that is immensely powerful for pilgrims viewing it today.

It is this kind of beauty and power that seduces me most in architecture of any era. It is the tectonic finesse and clarity of the Gothic cathedral that makes its forms and spaces so magical. It is the craft of its makers in material usage, construction, and detail that makes it so enduring. The forms of medieval piers and vaulting are so visually stunning that it is sometimes difficult to remember that their basis is technology. It is only when viewing very ordinary industrial buildings of this era like the modest forge at the Abbey of Fontenay that we realize that the origins of these forms were basic, practical construction methods. Technology is the seminal force in creating beauty.

It is these same traits in modern architecture that have been equally inspiring over the last century. The economy, complexity, and potency

03 Abbey of Fontenay, France
04 Farnsworth House, Plano, Illinois, Mies van der Rohe
05 La Tourette, near Lyon, France, Le Corbusier

of a small one-room building like the Farnsworth House by Mies van der Rohe is phenomenal and exquisitely beautiful. Construction becomes poetry. The simple, elegant steel frame lifted off the ground to clarify all its parts and connections is lean and direct. It is also deep, rich, and layered with interpretation. This tiny structure is solid, but also hovers. It is a frame, but also a volume. It is diminutive, but also monumental.

No less so, the economy, complexity, and potency of Le Corbusier's rugged concrete fortress at La Tourette are also moving, awesome, and very beautiful. Construction here is crude and elemental. Concrete is a heavy oozing material that denies precision, reveling instead in an intrinsic rawness. The result is concurrently modern and archaic. The technology is fresh and innovative, but also ancient and very basic.

In our own work we are constantly striving to learn lessons about materials, construction, and building technology that can inform the shape, texture, space, and character of our buildings. We pride ourselves in doing thorough research into whatever material we employ. Where will the stone be quarried? How does the color change in different veins of that particular geology? What kinds of equipment does the quarry employ? Can they polish, hone, or flame-finish on-site, or is that done elsewhere? Can the strength of the stone be utilized structurally? Can it be used as a structural composite with concrete so that stone becomes formwork, finish, and structural element? Can its thermal mass be employed to advantage? Who will lay the stone? What is the best kind of mortar joint to match the skill of the mason? How will the stone age over time? How will its moisture retention affect adjacent materials? Will it need to be cleaned or otherwise maintained?

But once the material is really known in terms of its molecular composition, its strength, its thermal qualities, its weight, its manufacture, its visual character, etc., many lessons are already apparent about design potential. It can be employed with truth and beauty or it can be applied superficially—just as visual effect. If one believes that craft is indeed important in the creation of beauty in architecture, then it is fundamental that architects must know materials, construction methods, and technologies, and use them with sincerity and mastery.

I am constantly impressed by the conscientiousness toward craft among great figures in any creative field. Outstanding clothing designers know fabrics deeply and intimately and depend on the material's weight, weave, drape, texture, etc. to create their magic. They know how to cut, fold, hang, and seam each piece to produce the desired effect. Great photographers know various films—their speed, their saturation, their stability, their ability to capture warm or cool tones. They know lenses, shutters, apertures, filters. Above all, they know light—its density, its brightness, its slant, and its volatility.

As architects we have a daunting quantity of technology that we must master. Far more than a clothing designer or a photographer,

06

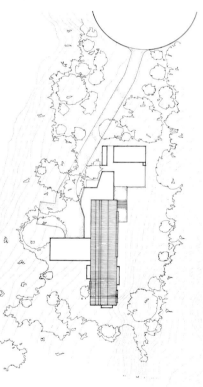

07

Concrete House, Austin, Texas
Concrete House—site plan

we must be committed to continual research and education to keep ourselves abreast of both longstanding and emerging technologies. Creativity depends not only on breaking new ground with new materials and assemblies, but also on using time-tested methods skillfully and with real freshness. In both of these arenas lies the potential to create truly beautiful architecture through the employment of great craft.

The Concrete House
The site for this 6,000-square-foot house had one of the most spectacular views in Austin. Situated on an east-facing hillside it captured long vistas of the Colorado River, Town Lake, downtown Austin, the Capitol Building, and the University of Texas tower. Lush vegetation shrouded the foreground so that almost no other houses intruded into the natural frame with the city beyond. But the site had a significant downside as well. It had very poor soil conditions—a deep layer of expansive clay on the surface with limestone bedrock a full 30 feet below. From the start, it was clear that a massive foundation with concrete piers shielded from the expansive clays would be required.

The technical challenge of counteracting the poor soil conditions of the site became a significant provocation in developing the character of the house. Because the piers were so expensive it was important to keep their number to a minimum. A simple rectangular volume for the main part of the house afforded an ideal geometry to accomplish this. In order to space the piers as far apart as possible, deep concrete beams were an obvious solution as a means to bridge between supports. But instead of thinking of the spanning members as just beams, we started to imagine them, because of their height, as walls. In fact, it seemed to make very good sense structurally to make the whole rectangular volume out of concrete resting on a minimum number of piers carrying the loads to bedrock. A simple, elegant structural solution to the thorny soil conditions became the initial gesture for the house.

The rectangular volume was stretched long and thin to align with contours on the site requiring as little cut-and-fill and disruption of the natural topography as possible. The contours of the land naturally aligned with the view downhill so that the long east-facing side of the volume would deliver forcefully one of the chief assets of the site. The thin dimension of the rectangle was sized so as to allow a clear span of floor trusses for the lower floor, similar trusses for the upper floor, and somewhat lighter and more shapely trusses for the roof.

In order to scale down the volume somewhat and as a means to achieve some functional flexibility in plan, wood appendages were "hung" off the concrete box. Literally cantilevered in most cases, they housed the kitchen, the entry, a balcony, a bay window, and a study space off the master bedroom. The most prominent of the wooden pieces softened the narrow end of the rectangle that faced the street. The small dimension, modest scale, and gentle material on this

08

09

10

08 Concrete House—entry
09 Concrete House—interior detail
10 Concrete House—west façade

approach side of the house combined with the preservation of most of the original site vegetation to render the house virtually invisible from the street.

The finish of the concrete was not forced to look clean, sleek, or consistent. The intention was to revel in the intrinsically imprecise nature of the material. It was important to have "bug holes" and even modest amounts of "honeycombing." Though care was taken to keep formwork clean and not to reuse it excessively, the texture of the plywood—even with an occasional "football"—was accepted as an appropriate reflection of the process of making. The resulting concrete finish is rich, warm, and variegated. The magic of liquid turning to stone, the heft and mass, as well as the handmade quality of the concrete, are palpable.

Both the thermal and moisture protection performance of the concrete and the process of putting it in place provided stimulating design challenges. Though any given section of the wall was produced in one pour, it was actually made up of two layers of concrete with an interstitial layer of styrofoam insulation between. The insulation was placed in the formwork between the steel reinforcing bars and held in place by metal discs while the concrete was poured around it. This resulted in an eight-inch-thick layer of concrete inside the insulation, lending remarkable thermal stability to interior spaces.

The other prominent materials of the house are used in a way that also seeks to reflect their physical properties. The wood volumes are "hung" in contrast to the firmly rooted concrete and are detailed to be light and thin. At the entry, the narrow wood siding bends to embrace a small gallery, demonstrating its greater flexibility and shape potential. The metal roof is demonstratively even more thin and bendable. It is a sleek membrane stretched over the concrete volume with contrasting precision and uniformity. All of the materials—concrete, wood, and metal—were intended to weather naturally, taking on a compatible patina of grays, buffs, taupes, and browns over time.

Inside, the generative structural approach delivers big open spaces compatible with the grand view and big sky outside. Though concrete walls can be a bit dominant if they fully enclose a small space, here they have plenty of room to breathe and feel nicely in scale with the space. The whole top level is like one big room. Even the master bedroom and bath on the south end bleed into the big living/dining spaces which occupy the rest of the floor. Bath and closet spaces are defined as pavilions floating in the larger room—their walls stopping far short of the ceiling. The master bedroom is more isolated, at the back, but is still spatially continuous at the top of the room through glass partitions.

The soft curve of the ceiling is delineated by asymmetrical bowstring trusses that gesture toward the view. The bottom chord of the trusses, which is structurally a tension member, is rendered as a thin rod of steel. The other chords are heftier wood reflecting their necessity

Concrete House—living room
Concrete House—fireplace
Concrete House—master bathroom

to carry compression and bending. The articulation of each structural material—concrete, steel, wood—playing its most appropriate role and reflecting its natural, physical traits, gives an almost anthropomorphic quality to the building. We can feel the strength and heft of some parts and the thinness and delicacy of others and palpably relate them to the bones and muscles of our bodies.

Floors in the big upper space are recycled longleaf pine. They are resawn in wide planks here after an earlier life in a warehouse in Louisiana. The rich, prominent grain of the pine gives it a strong, natural feeling that is very different from a polished hardwood floor. Cabinet work is made of painted ash, chosen because the ash grain subtly telegraphs through the painted surface. The fireplace is rendered cement—a slightly more refined version of the concrete walls. The hearth is made of steel that was sandblasted, burnished, and oiled to give it a patina like a gun barrel. Eased slightly away from the edge of the fireplace, the hearth does not convey heat and serves as a smooth bench for extra seating in the living area.

Several wall surfaces are made of Stramit, a very "green" material made from wheat stalks. Disposing of the refuse of wheat production is a big problem in Texas—especially with a ban on burning in many areas. The Stramit recycles this waste material by heating it under pressure until the natural glutens in the straw fuse the material together. The Concrete House used some of the very first sheets of Stramit produced in Texas by an experimental operation in the Texas Panhandle. The Stramit is layered together like butcher-block to produce a richly textured, honey-colored surface.

It has been fascinating to see the house embrace the eclectic set of possessions its owners have brought into it. A very wide range of art—from Old Masters to Jasper Johns—has graced the walls and looked remarkably at ease on them. The concrete has enough neutrality so as not to compete with art, but enough presence and tactility to place the art in a context that is real rather than self-effacing.

There is a powerful sense of craft in the Concrete House. It amply meets the three criteria set out by my law colleague. There is a rawness and economy in the house. It is not fussy or overwrought. It is simple and direct. There is also a complexity here. The dialogue between different materials and between the character of the materials, the character of the site, the character of the spaces within is layered and multifaceted. There are traces all over the house of how it was made, lending an extra richness of meaning. This is, as well, a house with real potency. It is a quiet house with a robust presence, a muscular house with genuine delicacy. Visitors walk away with vivid memories and striking impressions.

The beauty here comes from an exploration of craft and technology. It is not a sentimental or nostalgic beauty, but one rooted in a tactile, vivid experience of life. This beauty is like that of the bark of a tree or the fur of a cat (which both, after all, reflect their technological role). It is not about style or fashion, and it shouldn't need to be "understood." It is meant to just be palpable—able to be absorbed through the senses.

Wabash Parking Garage

It is perhaps perverse to even discuss beauty in the context of a building type so mundane as a parking garage. On the other hand, maybe finding (or making) beauty in the most humble of situations is really the significant challenge architects commonly face. Seldom does a client come to an architect wanting them simply to produce beauty. The agenda is substantially (and appropriately) to make a house to live in or an office building to work in or a school to learn in. The creation of beauty is generally several lines down on the list of wants, if it is there at all. The creation of beauty must be implicit in every project an architect approaches, and our means of creating beauty must be intrinsic in accomplishing other goals. That is why mining technology as a source of beauty can be so fruitful—even in a parking garage.

A parking garage is a wonderfully basic building type. It possesses very singular technology and very singular function. The Wabash Parking Garage accepts the very fundamental nature of itself as a building. It is made of five parts—a robust concrete frame, a delicate lightweight metal armature, translucent glass infill panels, open stairways, and a copper-clad elevator core. Each element is rendered clearly and explicitly in its own role and then married to the other parts as appropriate.

The concrete frame is made up of vertical piers and post-tensioned horizontal beams of the same width and about the same depth as the piers so that they work together visually as a system. The piers are turned with their long dimension perpendicular to the building for lateral stiffness and sit proud of the rest of the structure. Flat slabs span between the beams and are turned up at the perimeter to create curbs. A line marking the thickness of the slab itself telegraphs through as a reveal on the perimeter surface. Piers, beams, slabs, and curbs are articulated as individual, straightforward functional elements but are tied together by their common material—concrete.

A lightweight galvanized steel armature runs vertically between the piers and parallel to them, subdividing the short bays into three parts and the long bays into four parts. The armature is made of double steel angles bolted to each other and, in turn, bolted back into the concrete frame. The steel is clearly not meant to carry building loads, as it is hung off the structure itself. The armature is made denser at the bottom of the building and on the back by the addition of a metal trellis attached to the double angles. This trellis provides security as well as a framework on which jasmine vines are growing.

Galvanized metal clips attach the translucent glass panels to the steel armature. The glass is laminated with a milky film between the

14 Concrete House—east façade
15 Wabash Parking Garage, Austin, Texas— south façade
16 Wabash Parking Garage—outside corner detail
17 Wabash Parking Garage—pier detail
18 Wabash Parking Garage—stair and elevator towers

16

17

18

two layers giving it a rich luminous quality. It is set out a few inches from the concrete slab and straddles the slab and the opening above. This gives each piece of glass two distinctly different colors and different levels of light transmission. The glass has a slightly green tint which is more intense when there is a void behind it and has a more milky quality when it is backed by concrete. Its shimmery edges catch glints of light in the sun. The glass, which is an unusual material for a parking garage, allows maximum natural daylighting for the interior, fulfills guardrail requirements, functions as a visual screen for cars, and breaks down the utilitarian scale of the concrete structure.

Open stairways at two diagonal corners celebrate their difference in scale and geometry from the rest of the structure. With steps cantilevered off a central pier, they are dynamic and even jittery in comparison to the quiet regularity of the rest of the structure. Their pleasantness and sense of security encourages people to use them for vertical circulation rather than the elevators, adding another layer of animation. With simple cable rails and minimal enclosure, they capture views and breezes in every direction.

The elevator core anchors the building and marks its primary pedestrian entry and exit point. Sheathed in copper shingles with a rich, warm patina, this tall slender shaft is clearly all about vertical movement. Beside it, a cantilevered awning sheathed in galvanized metal with a clean, flush soffit marks the entry and provides shelter for shuttle users. Horizontal panels of the milky glass loosely define a pleasant outdoor lobby below it.

Lighting was a design issue dealt with thoroughly in this garage. Because strong light could have intruded upon the surrounding neighborhood, primary lighting is indirect—shining up onto the flat slabs and bouncing down into the spaces below. A bit of supplemental light at the perimeter of the building is provided by a series of small utility fixtures placed just inside the translucent glass. They light the edge of the building gently without flooding the area, and give a bit of sparkle to the building façade at night.

Though it had very low construction costs (less than a standard precast parking garage), this is a well-crafted building. It satisfies the criteria of economy, complexity, and potency. It is lean and simple, yet its assembly of clearly defined parts gives it richness and multiple readings. It is a striking object and has become something of a landmark on the heavily traveled street where it is located.

Addition to Seton Medical Center

Expressing the physical, technological properties of materials like concrete, stone, steel, or wood has, perhaps, a more obvious architectural manifestation than the expression of the qualities of a less tangible building material like glass. Often glass is used in buildings simply to create a void rather than to emphasize the inherent properties of the material itself.

Since the 1970s I have been haunted by Paul Scheerbart's little 1914 book *Glasarchitektur*, in which he extols the technological and aesthetic potential of glass to create new architectural possibilities. An original copy of the book from its first printing resides in the Humanities Research Center at University of Texas at Austin. The old and delicate appearance of the little text coupled with its progressive ideas that still seem fresh and new make a powerful impression. For Scheerbart, glass had a strong physicality with variable texture, thickness, color, light transmission, and even structural capacity. He was fascinated by emerging glass technologies of his era, and was enthralled by future potentials he could conceive for its application.

The material which is described with such passion in *Glasarchitektur* was imagined, even in its simplest application, as being configured in multiple layers with light between the layers. Scheerbart imagined "light columns" made of "light elements behind a completely glass surround" which would "not give the impression supporting." He envisioned such glass architecture would acquire "an almost floating quality." [17]

When we were called upon to add a large surgery center to an existing 1970s brick hospital with a rather heavy, dour visage, Scheerbart's ecstatic lightness came to mind. Part of the scope of the project was to create a new entry, a new chapel, and a new "front door" image for the whole hospital which would give a more positive, progressive impression. The aesthetic quality of glass and light used together offered the potential to create a sense of precision and technological sophistication alongside a certain softness and illusiveness. These were deemed appropriate for a building where modern medicine is concurrently dealing with both scientific exactness and human frailty and vulnerability.

As a functional device, a long thin volume faced with glass was strung along the north side of the new surgery center to bring light deep into rooms, which had to be below grade. The translucent glass draws in bright, cheerful light without sacrificing visual privacy for the recovery rooms inside. At night the glass volume is lit from within lending a safe, glowing presence to the medical center where activity occurs around the clock.

Though tied visually to the original hospital by low brick walls and by shared rhythms and proportions, the new building offers a bold new image for the medical center. It employs a serious investigation of glass technology to create a rich evocative beauty.

Austin Convention Center Expansion

Just eight years after the original 400,000-square-foot Austin Convention Center was completed, the success of the facility demanded a doubling in size. Though the original scheme had anticipated expansion, the site reserved for growth was only about half the size of the original site and was intended to accommodate only half as many

19 Addition to Seton Medical Center, Austin, Texas—north façade
20 Addition to Seton Medical Center—chapel interior
21 Addition to Seton Medical Center—façade detail

22

23

24

2 Austin Convention Center Expansion—model
3 Austin Convention Center Expansion—new Trinity Street façade
4 Austin Convention Center Expansion—glass pavilion at corner of Fourth and Trinity Streets

square feet as the initial building. The necessity to make the expansion twice as dense as the existing structure required a fundamental rethinking of the original scheme, though there was a strong desire to create a seamless whole (not an old and a new building) in the end.

There was still a very firm commitment, as well, to the urban design intentions of the original building. It was clearly important to maintain not only the variety and legibility of streetscape of the 1992 building, but also to retain its scale, which had been drawn from the best parts of nearby Congress Avenue and historic Sixth Street. This would be challenging given the much denser, and necessarily much taller, new building.

As in the original project, the loading area was located on the back side of the site, nearer the interstate highway and away from the action of downtown. The "big box" of the exhibition hall was placed in the center of the site, away from the important street edges. It was ringed with smaller, more active functions—lobbies, registration areas, and prefunction spaces, as well as stairways and escalators—addressing the street. Unlike the original building, a second layer was stacked on top of both the exhibition spaces and some of the peripheral areas. This level contained a 45,000-square-foot ballroom, extensive meeting room suites, and the lobbies and prefunction spaces required for them. The scale of the expansion from street level is kept similar to the original building by stepping the upper floor back, especially along the Trinity Street façade, in order to reduce its bulk. That provided the opportunity to create generous terraces off all upstairs prefunction spaces which have excellent views of downtown.

Another significant difference in the façade treatment between the two phases lies in the response of the peripheral elements to their varied surroundings. The first building was faceted on one edge to respond to the shape of the creek and was made partly of rough-faced limestone to match the character of the creek bed. The new building is firmly within the downtown grid, has great views of the skyline in two directions, and has a long face with north light available. It is, therefore, cleaner, crisper, lighter, brighter, and more open than the original.

Because expansion was anticipated to the north, the last element on that edge of the Trinity Street façade was made of an anodized aluminum skin—a material that could be easily matched later to provide a seamless transition between old and new. The joint between the two was originally intended to be a void which would be the terminus of Third Street looking east. (This would provide an interesting counterpoint to the iconic pavilion with its polygonal window which terminated the more active Second Street.) The void would be flanked by aluminum-faced volumes on either side—one old and one new. When the much larger program needed to be accommodated, footprint space became very precious, and it was not possible to leave a void. Instead, a large stair tower was placed between the metal pavilions and was faced in

25 Austin Convention Center Expansion—interior of glass pavilion
26 Austin Convention Center Expansion—interior detail

a very delicate woven stainless steel mesh. Because this treatment is so transparent and mysterious, the façade almost becomes a void, fulfilling the original intention.

Trinity Street, which is the "green finger" pedestrian link between Town Lake a block to the south and the Sixth Street Entertainment District two blocks to the north, has a somewhat tighter building façade along its edge than Fourth Street, which borders the site to the north. The Fourth Street façade is bolder in scale facing historic Brush Square (one of the four original public squares in the 1839 plan of the city) and a new high-rise convention hotel. Its grander "civic" scale acknowledges, as well, the future role of Fourth Street as the city's major east/west transit corridor. At the corner of Trinity and Fourth streets is an elegant glass pavilion which creates a new "front door" for the entire complex. Its presence on this very visible corner, anchoring Brush Square, establishes a powerful landmark for the convention center in the downtown fabric. The pavilion contains a single space 90 feet tall, 62 feet wide, and 142 feet long which serves as a critical vertical link between exhibition halls on the ground floor and ballroom/meeting room spaces above.

In this pavilion the notion of technology as a source of beauty is particularly apparent. Both inside and outside the art and the craft of building are explored in every surface and detail. It is the "jewel" of the complex with views from all over the building on both levels terminating in its light, airy volume. It is a powerful orientation device in a large, complex building as well as a prestigious venue for receptions and other significant occasions. It is important that it be beautiful.

Rising through the dramatic space is a striking escalator with its two legs joined by a landing that cantilevers out through the west façade. The escalator turns 160 degrees at the landing to ascend its full 54-foot height at the prefunction space outside the upper-level ballroom. The escalator ride provides a vertical promenade through the space and through the city with spectacular views of the skyline and the Hill Country landscape beyond. At night the pavilion becomes a beacon for the convention center with the soft light reflected off its angular ceiling providing a warm glow through its crystalline façade.

From an urban design point of view, from considerations of orientation and function, from an experiential point of view and in terms of daylighting performance, it was critical that this strategic corner pavilion be as transparent as possible. Addressing one of the most important open spaces in the city, the building's long north façade gave a rare opportunity for openness without compromising thermal performance. The technologies developed here came as a means to promote transparency in an extraordinary way—one that would dramatically reinforce broader design intentions.

The outside "wall" of the pavilion is actually a ten-foot-wide zone which deals interdependently with structure and enclosure. The

27 Austin Convention Center Expansion—interior detail
28 Austin Convention Center Expansion—west face of glass pavilion

structure needed to be kept as slender as possible in the direction of the view and yet, because of its great height, with no intermediate floors for bracing, it had to exercise some means of achieving stiffness. Within the ten-foot zone are a series of eccentrically proportioned piers (long and very thin), a five-foot-wide gap (filled with cable and rod bracing), and a curtainwall (literally hung from the roof) made of shingled glass. The piers, the gap, and the curtainwall cooperate to carry gravity loads, resist lateral loads, create a thermal and moisture barrier, diffuse natural light, and provide an extraordinary transparency in the direction of the view.

The piers are made up of three components. The lower 18 feet of each pier is a reinforced concrete shaft 16 inches wide and four feet deep. The center 60 feet is a slender steel box column six inches wide and 30 inches deep. The top 12 feet consists of a "tree" made of four solid steel bars three inches in diameter arranged in an inverted pyramid. The composite pier is an elegant essay on the forces and potential inherent at each stage along the pier's length and the capabilities of steel and concrete in various configurations to deal with those forces.

The glass curtainwall minimizes the size of structural supports required by reducing compressive stress and compressive buckling. Vertical structural members are kept in tension by hanging the façade from the roof beams which are ten-foot-on-center atop the steel "trees." T-shaped hangers drop 72 feet from each beam with long slender blades—3/4-inch by 18 inches—facing outward. The glass is placed so as to slope against the depth of the blades in 12-foot tiers creating a shingled effect. The portion of the blades outside the glass becomes an elegant saw-tooth fin of varying depth. Between the Ts, horizontal members eight inches wide and four inches tall on 12-foot vertical centers create supports for the five-foot by 12-foot, one-inch-thick glass panels. (The horizontal members also transfer wind loads from the curtainwall back to the box column via the aforementioned struts and cables.) Each glass panel is made of two layers of low-emissivity glass with a ceramic frit on the inside of one sheet to improve thermal performance and reduce glare.

The "tour-de-force" of the pavilion is a large glass screen wall on its west face. Though most of the surface of the pavilion needs no special sun protection because it faces north and is made of high performance glass, the glazing on the west required special attention. The glass screen provides sun shading, but it also becomes a piece of glass sculpture and a solar energy farm. A series of vertical panels with photovoltaic cells faces southwest to capture maximum exposure for solar collection and to filter the sun during the hottest part of the day. A series of cobalt blue panels with an acid etched surface faces due west and contribute color and animation while further reducing heat and glare on the interior. The structure supporting the glass screen is a mirror image of the pier structure of the pavilion. Both piers and screen

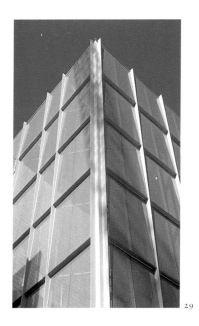

29 Austin Convention Center Expansion—corner detail of glass pavilion
30 Austin Convention Center Expansion—district plan

project 12 feet above the pavilion to fully block the summer trajectory of the sun.

Design of the pavilion was a collaborative effort involving many consultants as well as the architectural team. James Carpenter, a glass sculptor, contributed his artistry on the west screen wall. Matt King, an engineer with Ove Arup Partners consulted on structural frame concepts. Chuck Naeve and his firm, Architectural Engineers Collaborative, elaborated and executed the structural design. Davidson Norris consulted on daylighting and glass selection. Ann Kale contributed to the lighting design. Beck Steel fabricated and erected the primary structure. Win-Con installed the glazing. Austin Energy funded and assisted with design of the photovoltaics. Many others were involved in fundamental ways as well.

The technology and the craft that was achieved collaboratively here at a high level of sophistication reflects the same three qualities observed in other projects—economy, complexity, and potency. The spidery delicacy of the structure represents an effort to use the very smallest quantities of steel to perform any given task. The minimalist quality of the struts, cables, and "trees," in particular, represents an extraordinary thrift and efficiency. It is almost by magic that they do their job. An elegant aesthetic economy pervades the whole space—from structure to glass to wall treatments to handrails.

A rich complexity likewise permeates the place. The intricate combination of parts is inextricably linked into a coherent whole. Structure and enclosure are individually articulated, but interdependent. A constant dialogue is struck between issues of construction, light, view, thermal performance, and functional accommodation. Urban demands and interior needs codetermine shapes, forms, and spatial qualities. Everything is interrelated.

The result is a strong sense of place, a real architectural potency. This is a landmark object in the city and a jaw-dropping room inside. It introduces a series of spaces on the interior of the convention center that range from yeoman to ethereal depending on their role. (Exhibition Halls are workhorses. The sexy "light blades" in one reception hall are hummingbirds.) But this front door pavilion is the powerful glue that holds the whole together.

The beauty here is intricate and fascinating—like the works of a watch or the pattern of dew on a spider's web. There is something sure and inevitable about it, but also an element of quirkiness and individuality. There are no precious materials like marble or rosewood or gold leaf—sometimes associated with beauty. But there is a fineness that comes from an investment of human attention to thinking, conceiving, and making.

A

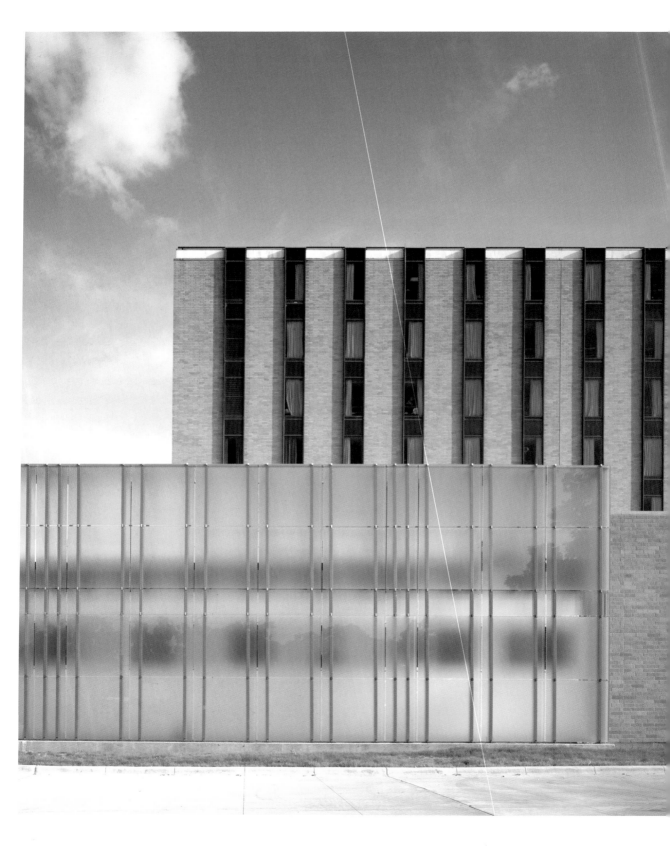

Previous page: **A.** The aesthetic qualities of glass and light used together create a sense of precision and technological sophistication alongside a certain softness and illusiveness–both qualities deemed appropriate for a building where modern medicine is dealing concurrently with both scientific exactness and human vulnerability.

B. A long, thin volume faced with glass was strung along the north side of the new surgery center to bring light deep into rooms which had to be below grade. **C.** Two kinds of translucent glass veil the structure without obscuring it and form a luscious, luminous skin for the building. **D.** Though tied visually to the original hospital by low brick walls and shared rhythms and proportions, the building offers a bold new image for the medical center.

C

E. Luminous glass continues in the new main lobby of the medical center, blending natural and artificial light. F. New chapel is a serene jewel box with a light, bright, positive ambiance. G. Suites of surgery recovery rooms provide a very soft, diffuse natural light for the benefit of both patients and staff. H. Primary waiting room has generous north light and views to mature trees saved in the planning of the expansion. I. Glass-faced canopy works with mature trees to provide a fresh new image at the "front door" of the medical center.

J **K**

J. Because it is a 24-hour environment, the image of the building at night took on special importance. **K.** After dark, the back-lit glass provides a soft glow which gives a strong sense of both welcome and security.

132 AUSTIN CONVENTION CENTER EXPANSION

A

B

C

Previous page: **A.** The notion of technology as a source of beauty is explored via the craft of the building with materials and detail, providing an impetus for architectural character.

B. The scale of the expansion from street level is kept similar to the original building by stepping the upper floor back, especially along the Trinity Street façade, in order to reduce its bulk. **C.** Firmly within the downtown grid with great views to the downtown skyline, the new building is cleaner, crisper, lighter, brighter, and more open than the original. **D.** The glass curtainwall on the north face of the pavilion on the corner of Fourth and Trinity Streets minimizes the size of structural supports by using thin steel blades in tension. **E.** Glass is placed so as to slope against the depth of the blades creating a shingled effect that catches the changing light and color of the sky in a dramatic way.

E

F

G

F. Massive quantities of egress stairs are used to animate what would otherwise be a blank wall on the east face of the expansion. G. The glass screen on the west face of the corner pavillion provides sun shading, but also becomes a piece of glass sculpture and a solar energy farm. H. Rising through the dramatic space inside is a striking escalator with its two legs joined by a landing that cantilevers out through the west façade.

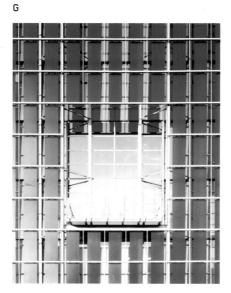

H

I

J

I. Light is introduced to an interior second floor pre-function space via skylights coupled with a curved reflective wall and a milky glass screen. J. Glowing walls that result diffuse the bright Texas sun as it animates opposing walls in the pre-function space. K. The upper level lobby off the ballroom offers expansive vies of downtown Austin through the delicate glass curtainwall. L. West sun is filtered through translucent blue glass as well as angled photovoltaic panels. M. There is something sure and inevitable about the structure and technology of the building, but also an element of quirkiness and individuality.

L M

END NOTES CREDITS

1. Louis H. Sullivan, *Kindergarten Chats and Other Writings* (New York: Dover Publications Inc, 1979), 64.
2. Ibid., 65.
3. Ibid., 99.
4. Ibid., 28.
5. Ibid., 32.
6. Frank Lloyd Wright, *The Natural House* (New York: Bramball House, 1954), 15, 17.
7. Thomas Beeby, *Modulus* (Charlottesville: Journal of the University of Virginia, Spring 1981).
8. Goran Schildt, *Alvar Aalto* (New York: Rizzoli, 1984) 102.
9. Paul Rudolph, "Rudolph," *Perspecta: The Yale Architectural Journal* 7 (1961): 51.
10. Ibid.
11. Simon Grey and Graham Farmer, "Reinterpreting Sustainable Architecture: The Place of Technology", *Journal of Architectural Education* 54/3 (February 2001): 146.
12. Ibid., 144.
13. Stanford Anderson, "The Convention Center and the City", *Progressive Architecture* (October 1992): 106.
14. Ibid.
15. Joel Barna, "Convention Community", *Progressive Architecture* (October 1992): 108.
16. Elaine Scarry, *On Beauty and Being Just* (Princeton: Princeton University Press, 1999), 52.
17. Paul Scheerbart and Bruno Taut, *Glass Architecture and Alpine Architecture* ed. Dennis Sharp (New York: Praeger, 1972) 41, 46, 50.

AUSTIN CONVENTION CENTER (ORIGINAL PROJECT)
AUSTIN, TEXAS 1992

Project Team
The Austin Collaborative Venture
Prime A/E and Managing Entity
PageSoutherlandPage
Principal-in-Charge
Matthew F. Kreisle, III, AIA
Design Principal
Lawrence W. Speck, FAIA
Project Manager
Charles L. Tilley, AIA
Architectural Team
PageSoutherlandPage:
Ham Frederick, Richard Perkins, Tim Cuppett, Dan Barrick; Cotera Kolar & Negrete (formerly VCK): Art Arrendondo
Project Management
Gilbane Building Company
Structural Engineering
Ellerbe Becket
Civil Engineering
PageSoutherlandPage: Jim Alvis, PE; Tom Burson, PE; Richard Hodges
Transporation Engineering
Wilbur Smith
On-Site Representative
PageSoutherlandPage:
Tom Golson; Cotera Kolar & Negrete: Art Arrendondo
Interior Design
Cheryl White, IIDA
Archeological Survey
Hicks & Company
Audio / Visual and Acoustics
Boner Associates
Communications
OTM Engineering
Fire Protection Engineering
Rolf Jensen
Food Service
H. G. Rice
Food Service Programming Services
William Caruso & Associates
Geotechnical
TETCO
Topographical Engineering
Carmelo L. Macias
MEP Noise Control Design
Jack Evans & Associates
Graphics & Signage
Fuller Dyal & Stamper
(now known as fd2s, inc.)
Landscape
Johnson Johnson & Roy
Lighting
Archillume Lighting Design
Roofing
Jim Cavalier
Security
Schiff & Associates (now known as Kroll Security Services Group)
Owner
City of Austin: City of Austin Management Services, Department of Public Works and Transportation, NathanSchneider AIA, Architect Contract Division, Fred Evins, Architect
Photography
Richard Payne, FAIA; Blackmon Winters Incorporated Architectural Photographers

AUSTIN CONVENTION CENTER (EXPANSION)
AUSTIN, TEXAS 2002

Project Team
The Austin Collaborative Venture
Architect
Austin Collaborative Venture:
PageSoutherlandPage
Principal-in-Charge
Matthew F. Kreisle, III, AIA
Design Principal
Lawrence W. Speck, FAIA
Project Designer
Brett Rhode, AIA
Project Managers
Charles L. Tilley, AIA and
Brett Rhode, AIA
Project Architect
Ken McMinn, AIA
Architecture
Cotera Kolar Negrete & Reed Architects, Austin, Texas; Limbacher and Godfrey Architects, Austin, Texas
Project Management
Gilbane Building Company
Structural Engineering
Architectural Engineers Collaborative, PLLC; Arup, Inc. (Atrium Structural Concept)
MEP/Civil Engineering
PageSoutherlandPage:
Robert F. Zelsman, PE (Engineering Coordinator), David Ashton, PE (Electrical Engineering), Robert Burke, PE (Plumbing and Fire Protection)
Contract Administration
Robert Hill, RA
Interior Design
Cheryl White, IIDA
Glass / Glazing / Photovoltaic Design
James Carpenter Design Associates, Inc.
Daylighting Design
Carpenter / Norris Consulting, Inc.
Lighting Design
Ann Kale Associates, Inc.
General Contractor
Spaw Glass Contractors
Owner
City of Austin: Austin Convention Center Department; Robert Hodge, Director
Photography
Tim Griffith

BARBARA JORDAN PASSENGER TERMINAL AUSTIN-BERGSTROM INTERNATIONAL AIRPORT
AUSTIN, TEXAS 1999

Architect
PageSoutherlandPage
Principal-in-Charge
Matthew F. Kreisle, III, AIA
Designer
Lawrence W. Speck, FAIA
Associate Architect
Gensler
Designer
Ron Steinert, AIA
Airport Planner
Landrum & Brown:
Gary Blankenship, AIA
Project Manager
Charles L. Tilley AIA
Project Architect
Tom Cestarte, RA
Architecture
Cotera Kolar Negrete & Reed, Architects; Architecture + Plus (Signage & Graphics); BLGY, Inc.
Structural Engineering
Jaster-Quinanilla & Associates, Inc.
MEP/Civil Engineering
PageSoutherlandPage: Robert F. Zelsman, PE (Engineering Coordinator), David Ashton, PE (Electrical Engineering), Robert E. Burke, PE (Plumbing and Fire Protection Engineering) Jim Peery, PE (HVAC Engineering); KLW Engineering

(Electrical Engineering); Burns & McDonnell Engineering Company, Inc (Electrical 400HZ Engineering); Rolf Jensen & Associates (Fire Protection/Life Safety Engineering)
Cost Estimating
Hanscomb, Inc.
(Now Hanscomb Faithful & Gould)
Interior Design
Cheryl White, IIDA (Lead Designer); Wendy Dunnam Tita, Associate AIA (Designer)
Recycling
H.D.R. Engineering, Inc.
Acoustics
Boner Associates, Inc.
Baggage Handling
URS Greiner Aviation Services Group, Inc.
Roofing Consultant
Jim Cavalier and Associates
General Contractor
The Morganti Group
Owner
City of Austin, Department of Aviation: Charles W. Gates, Director of Finance and Administration
Photography
Paul Bardagjy; John Edward Linden

COMPUTER SCIENCES CORPORATION FINANCIAL SERVICES HEADQUARTERS
AUSTIN, TEXAS 2002

Architect
PageSoutherlandPage
Principal-In-Charge
Matthew F. Kreisle III
Lead Designer
Lawrence W. Speck, FAIA
Architecture Quality Assurance
Jane M. Stansfeld, FAIA
Project Management
E. Douglas McClain, PE; F. Keith Hall, AIA
Project Architect
Gwen Jewiss, AIA
Project Architect (Parking Garage)
Peter Hoffmann, AIA
Structural Engineering
PageSoutherlandPage:
William N. Berezovytch, PE;
Architectural Engineers Collaborative:
G. Charles Naeve, PE;
Jaster-Quintanilla & Associates, Inc., (for Contract Administration)
Tom Durham, PE
MEP Engineering
PageSoutherlandPage:
Robert F. Zelsman, PE (Engineering Coordinator), David Ashton, PE (Electrical Engineering), Robert Burke, PE (Plumbing and Fire Protection), Andy Baxter, PE (HVAC Engineering)
Interior Design
PageSoutherlandPage; Jen Bussinger, IIDA; Bob Stapleton Associate, IIDA; Reheannon Cunningham, IIDA; Wendy Dunnam Tita, Associate AIA
Design
Tom Hurt; Rommel Sulin; Scott Grubb, Associate AIA; Andres Cuerto
Lighting Designer
Ann Kale Associates, Inc., New York, NY
Acoustician
Dickensheets Design Associates, Austin, TX
General Contractor
Hensel Phelps Construction Company, Austin, TX
Client
Computer Sciences Corporation
Photography
Paul Bardagjy; Tim Griffith

ROBERT E. JOHNSON LEGISLATIVE OFFICE BUILDING
AUSTIN, TEXAS 2000

Architect
PageSoutherlandPage
Principal-in-Charge
Matthew F. Kreisle, III, AIA
Design Principal
Lawrence W. Speck, FAIA
Project Management
Charles L. Tilley, AIA
Project Architect / Contract Administration
Robert Hill, RA
Associate Architect
Cotera Kolar Negrete & Reed
Structural Engineering
Jaster-Quinanilla & Associates
MEP Engineering
PageSoutherlandPage:
David Ashton, PE (Electrical Engineering), Robert Burke, PE (Plumbing and Fire Protection); James Peery, PE (HVAC Engineering)
Designer
Wendy Dunnam Tita, Associate AIA
Landscape Architecture
The Landscape Collaborative
Sustainability Consultants
Berkebile Nelson Immerrschuh McDowell
Sustainability
NSAR Group
Cost Estimating
Project Cost Resources
Security System
Schiff & Associates
Contractor
SpawGlass Contractors, Inc.
Client
General Services Commission, State of Texas
Photography
Paul Bardagjy

ROUGH CREEK LODGE AND CONFERENCE CENTER
GLEN ROSE, TEXAS 1998

Architect
PageSoutherlandPage
Project Principals
Matt F. Kreisle, III, AIA; Mattia J. Flabiano, III, AIA
Lead Designer
Lawrence W. Speck, FAIA
Project Manager
Tom Cestarte
Project Architect
Gwen Jewiss, AIA
Architectural Team
Tim Cuppett, RA; Brett Rhode, AIA; JoAnn Lenyo; Donovan Oliff; Regina Rawner; David Boren; Barbara Ellison; James Murff
Structural Engineer
Jaster-Quintanilla & Associates, Inc.: Chuck Naeve, PE; Richard Martin
Civil Engineer
Childress Engineers: Benjamin S. Shankin, PE; Robert T. Childress, III, PE
MEP Engineering
PageSoutherlandPage:
David Ashton, PE (Electrical

ngineering); Robert Burke, PE
Plumbing and Fire Protection);
ames Peery, PE (HVAC Engineering)
MEP Contract Administration
David "Fritz" Schuetzeberg
Interior Design
Vivian Nichols Associates:
Diedre Duchene, PIC; Wayne Lambdin,
PM; Debbie McKee, IPA; Ashley Slagle,
PD; Kris Walsh; Nancy McMahon;
ennifer Griffin
Lighting Design
Craig Roberts Associates
Pool Design
Armstrong Berger, Inc.:
John H. Armstrong
Landscape Architecture
Site Planning Site Development (SPSD):
Cliff Mycoskie
Construction Documents
David Boren
Contract Administration
Bob May
Model
Mike Smith; Ryan Coover;
Lotte Stavenhagen
Photography of model
Geno Esponda
Food Service Consultants
H.G. Rice and Company:
Don Kohne
General Contractor
Thomas S. Byrne, Inc.: John Availa, Jr.
(COB); Mike Patterson (VP);
Elias N. Najjar (VP of Estimating)
Gary Steele
Owner
J. Q. Enterprises:
John Q. Adams, Sr. (COB)
Photography
R. Greg Hursley

AUSTIN CITY LOFTS
AUSTIN, TEXAS 2004
Architect
PageSoutherlandPage
Prime A/E and Managing Entity
PageSoutherlandPage
Principal-in-Charge
Matthew F. Kreisle, III;
Jane Stansfeld
Lead Designer
Lawrence W. Speck
Project Manager
Brett Rhode
Design Team
Brett Rhode, AIA; Ken McMinn, AIA;
Ronn Basquett; Jen Bussinger, IIDA;
Brian Cady; Rheannon Cunningham,
IIDA; Cesar Esclante; Elizabeth
Fitzpatrick; Randy Gains; Haia Ghalib;
Flora Gui; Bryan Haywood, RA; Chad
Johnson; Maire Katya; Alice Kramr;
Katherine McPhail; Robert Schacher;
Sylvan Schurwanz; Tanya Schmidt;
Ricardo Solis, RA; Roberta Swischuk
Specifications
Troy Templin, AIA
Structural Engineering
Architectural Engineers Collaborative:
Charles Naeve
Civil Engineering
Bury and Partners
MEP Engineering
Tom Green and Company
(MEP-SD/DD);
Fox Mechanical, Inc. (M-CD);
Johnson Consulting Engineers (EP-CD)
Interior Design
Henderson Design Group:
Marla Henderson
Landscape Architects
Big Red Sun
Contract Administration
Troy Templin, AIA
Cost Estimating
Faulkner Construction Co.
General Contractor
Rogers-O'Brien Construction Company
(Faulkner Construction Company)
Client
CLB Partners: Ron Cibulka
Photography
Tim Griffith Photographer

ADDITION TO SETON
MEDICAL CENTER
AUSTIN, TEXAS 2005
Architect
PageSoutherlandPage
Prime A/E and Managing Entity
PageSoutherlandPage
Principal-in-Charge
Matthew F. Kreisle, III
Lead Designer
Lawrence W. Speck FAIA
Project Manager
Doug McClain, PE

Design Team
Jay L. Willmann, AIA; Kurt Neubek,
FAIA; Kregg Elsass, AIA; Peter
Hoffmann, AIA; Chad Johnson;
Jerry Segner, RA; Jeff Jewesson, AIA;
James Murff, RA; Bryan Haywood, RA;
Scott Grubb, RA; Karla Jackson;
Kelly Akers, IIDA
Structural Engineering
Datum Engineers, Inc.:
Marty Sloan, PE
Civil Engineering
PageSoutherlandPage: James Alvis, PE;
Judd Willmann, PE
MEP Engineering
Smith Seckman Reid, Inc.
Transportation Engineering
WHM Transportation
Engineering Consultants, Inc.
Furnishings
Healthcare Interiors, Inc.
Signage / Graphics
fd2s, inc.
Landscape Architects
TBG Parnters
Medical Planning
The Innova Group
Medical Equipment
Parallel Solutions, Inc.
General Contractor
Vaughn Construction
Client
Seton Medical Center
Photography
Tim Griffith Photographer

Copyright © 2006 By Edizioni Press, Inc. All rights reserved. No part of this book may be reproduced in any form without written permission of the copyright owners. All images in this book have been reproduced with the consent of the artists concerned and no responsibility is accepted by producer, publisher, or printer for any infringement of copyright or otherwise, arising from the contents of this publication. Every effort has been made to ensure that credits comply with information supplied.

First published in the United States of America by Edizioni Press, Inc.
469 West 21st Street New York, New York 10011
www.edizionipress.com

ISBN: 1-931536-35-X

Library of Congress Catalogue Card Number: 2004116798

Printed in China

Design: Perrin Studio
Editor: Sarah Palmer
Editorial Assistant: Nancy Sul